Rapid Prototyping of Software for Avionics Systems

FOCUS SERIES

Series Editor Guy Pujolle

Rapid Prototyping of Software for Avionics Systems

Model-oriented Approaches for Complex Systems Certification

Nicolas Larrieu
Antoine Varet

ISTE

WILEY

First published 2014 in Great Britain and the United States by ISTE Ltd and John Wiley & Sons, Inc.

ISTE Ltd
27-37 St George's Road
London SW19 4EU
UK

www.iste.co.uk

John Wiley & Sons, Inc.
111 River Street
Hoboken, NJ 07030
USA

www.wiley.com

Library of Congress Control Number: 2014949836

British Library Cataloguing-in-Publication Data
A CIP record for this book is available from the British Library
ISSN 2051-2481 (Print)
ISSN 2051-249X (Online)
ISBN 978-1-84821-764-5

Contents

Introduction

Aeronautics[1] has for some time been stressing the need for safety systems that prevent the malicious use of materials or services that companies install onboard airplanes for their own use and requirements.

With the future arrival of onboard Internet access in most cabins, the need for safety will become a priority. Currently, no safety[2] solution exists that, on the one hand, manages this new type of air-to-ground traffic (belonging to the *Aeronautical Passenger Communication* [APC] family) and, on the other hand, is capable of integrating it with other types of traffic between the aircraft and ground (*Aeronautical Operational Control* [AOC], *Aeronautical Administrative Communication* [AAC] and *Air Traffic Services Communication* [ATSC] traffic, for instance), while maximizing the level of strength offered.

The majority of approaches to "air" safety focus on methods and techniques that make safe exchanges within the airplane or a link used exclusively for communicating

1 Aeronautics designates the sciences and technologies that aim to construct and develop an aircraft in the Earth's atmosphere.

2 In the context of air travel, safety is the property of innocuity in a system. It aims to protect the system from failure and breakdown.

with air control. The Avionics Full-Duplex Switched Ethernet (AFDX) onboard avionic[3] network cannot, therefore, be physically accessed by passengers: this segregation ensures its security[4]. Meanwhile, the Air Traffic Communications (ATC) air-to-ground flow uses the frequencies dedicated to civil aeronautical communication, which are prohibited by law for any other use.

Though necessary, this safety problem will no longer suffice when the avionic network interconnects with the remaining communication networks, including the Internet network. It is becoming increasingly apparent that this is an unavoidable step. Indeed passengers – who are used to being able to connect to the Internet everywhere on the ground – are increasingly calling for access to their work tools from their cabin seat.

This book will, therefore, put forward a software engineering method that enables the development of a security architecture for the set of aeronautical communications that will complement the safety architecture used in the avionic network and that will, moreover, facilitate a safe interconnection between the world of the "airplane" and the external world (for instance, the Internet network).

The architectural solution put forward in this book is principally based on a central routing, filtering and safety component for the flow of aeronautical data. The model-based software engineering method presented in this book

3 Avionics is the set of electronic, electric and computer equipment in an aircraft. Thus, it is a subset of the aeronautical domain: avionics exclusively concerns the inside of the airplane, whereas aeronautics encompasses the avionic domain in addition to its environment, including control and navigation facilities on the ground.

4 In the context of air travel, security is the property of immunity in a system. It designates a system's capacity to manage external threats and dangers.

enabled research into design and development to be realized. This led to the central avionic component called the Secure Next Generation (SNG) router being developed. The functionalities implemented by the SNG router are summarized in Figure I.1.

One of the challenges of this work was to take into account the standardization issues adopted by the avionic world, to integrate them as prerequisites and to propose an overall safety system involving the airplane segment, the ground-board segment and the ground segment.

Indeed, this work can only be scientifically and industrially sustainable if it takes into account the set of security and safety criteria characterizing the different environments and considers, from the offset, the different principles of standardization for the latter.

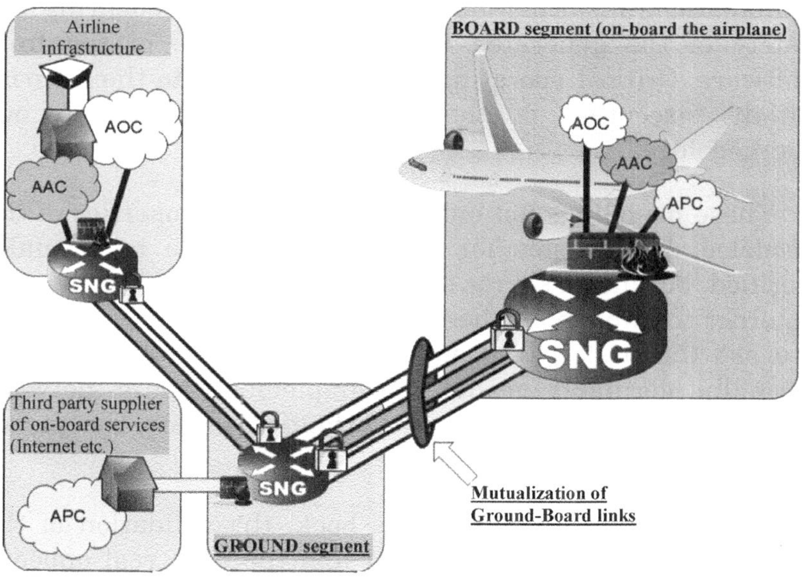

Figure I.1. *Example of the deployment of the SNG router in the aeronautical context*

I.1. The use of model-based approaches

The complexity of developing onboard avionic systems led us to preliminary research into the development process, which then led to the creation of a generic methodology of design. This new methodology for the development of avionic software and rapid prototyping takes into account, from the moment the scope statement is established, the security and safety constraints associated with the system to be implemented.

Our development methodology for the SNG router is based on functionality modeling, the automatic transformation of models in source code and the use of critical operating systems that implement virtualization. Indeed, the models have undeniable advantages in terms of the speed and reusability of the development, as well as for the verification and validation of the modeled systems. Automatic transformation contributes toward ensuring the safety of the generated code, and therefore of the final software. Critical operating systems are keys to the security of the execution of the software generated with our methodology.

These modelings not only enable certain properties of the modeled mechanisms for the SNG router to be formally verified, but also a source code to be automatically generated in order to implement these mechanisms. To our knowledge, we are the first in the aeronautical domain to have used formally verifiable models to implement these network mechanisms via a model transformation toward the source code.

Discussed at the end of this book, the validation of the router software on the emulator and then the real onboard system as well as the quantitative and qualitative evaluation of the network performance of this router have thus enabled the cycle of industrial development – on which this research is based – to be completed and finalized.

It is essential that from this introduction the readers understand the importance of model-based development for the design of complex computer systems. This importance translates into considering the possibility of using software tool chains in certificate procedures for aeronautical systems (see documents RTCA DO-178C and DO-331 that will be presented in this work), enabling software source code to be generated from high-level models, which will be deployed in critical aeronautical systems.

I.2. Structure of this book

In order to understand the interest of model-based approaches, it is important to place them in wider context than that of the aeronautical domain (this is the focus of Chapter 1). Chapter 2 will then discuss the specificities of the domain of the design of aeronautical and avionic systems, and an original model-based design method will be presented. This method was designed to meet the particular needs of aeronautics. This chapter will detail the methodological contribution of this work, which proposes a new method of model-based rapid prototyping. Chapter 3 will detail the different stages involved in the aforementioned method. This method will be illustrated with a concrete example of the development of a critical onboard system: a secure and safe router for new-generation avionics (SNG router). This will be the focus of Chapter 3. This represents a concrete example of the complex onboard system in the avionic domain. The specificities of aeronautics, our domain of interest for this software engineering task, will be introduced. We will stress the need for new means of interconnection as well as the requirement that both the security and safety of the communication be considered. We will illustrate, in particular, how this methodology can be instantiated for the specific case of the development of a safe onboard router. The chain of tools used to apply our methodology will be detailed. Examples will be given for each tool. This final chapter will conclude with a summary of the performance of the system, which was developed via the intermediary of our model-based methodology.

The benefits and advantages of our methodology for design and development as well as for the overall performance of the system produced will thus become apparent. From the results, we are able to state that the methodology presented in this book has a high level of effectiveness.

1

Developing Model-Based Design Methods in Software Engineering

The objective of Chapter 1 is to give an overview of the different model-based design methods in the domain of software engineering and of their development over the past three decades. We have decided to pause for a moment on the most commonly used language for writing models in the domain of system design today: *Unified Modeling Language* (UML). This standardized language ISO/IEC 19501:2005 [ISO 01] abstractly describes any system whose aim is to be applied via the implementation of a programming language (paying no attention to the specific programming language from the offset, and therefore not restricting itself to the limits of the expressiveness of this programming language). Once the important UML modeling concepts have been presented, the various advantages of this type of approach will be considered: modularity, reusability, portability and so on. In section 1.3.4, we will discuss system validation. Indeed, using design models coupled with formal validation methods for these models is beneficial in terms of speed and strength with regard to the certification of the final system.

1.1. The history of model-based design

From its beginnings in the 1960s, computer program design was unguided and left to the discretion of development teams. This free approach met the design needs of simple programs that had to be run on extremely hard to use hardware systems. Intellectual effort, therefore, principally concerned "how" to make the underlying electronics compute what developers wanted it to calculate. Thus, assembly languages were king.

However, university mathematicians were working on new and complementary approaches that considered real machines indeterminate elements; programs were first and foremost considered abstract, and they were then concretized manually or with tools called assemblers, linkers and compilers. The wide variety of possible uses foreseen for computer-based systems resulted in the development of a large number of programming languages. In 1977, Professor M. Halstead counted some 1,000 different languages [HAL 77] and excluded the countless variations and versions of each of these languages.

Though the object of endless disputes between enthusiasts, these two approaches – university and industry – defended, on the one hand, by mathematicians and, on the other hand, by electronic engineers, have always been complementary. The abstraction of high-level languages has, therefore, democratized these new computer systems whose internal operations are proving to be extremely complex. In return, improvements to electronics have influenced the formalism of languages, their opportunities and performance.

A new scientific era began in the late 1970s: the era of models. In fact, the exponential complexification of computer systems quickly limited humans' ability to design programs. The introduction of new and more formal methods guiding design enabled these limits to be stretched. Thus, the 1980s saw the appearance (or rather the reappearance and the adaptation) of graphic and text languages. These languages describe systems in a formal way and, instead of being used by machines, coordinate the different participants in application development.

In this new pattern of work, developers together establish a contractual framework for the program, its objectives and the principles of its internal operations. These principles are translated into lines of code in one or several appropriate languages and are finally compiled in a binary form that the machine understands and executes.

Different formalisms have been tested according to needs. Some are adapted to describe the structure of a computer program or of data, others to expose the operation state of the program at a given moment, others still to describe the succession of states through which the program passes or can pass. In the mid-1990s, three major object-oriented design methods were combined under the aegis of the *Object Management Group* and given the name Unified Modeling Language (*UML*). The UML is the focus of the following section.

1.2. *The Unified Modeling Language*, a support for model-based methods

1.2.1. *The philosophy and history of the Unified Modeling Language*

The principle of using classes and objects rather than functions to model a computer program is relatively old: the programming language Simula [POO 87] introduced this concept to the computer world from 1967. In 1976, the programming language Smalltalk [BRI 96] built on this concept by generalizing its semantics.

However, from 1983, the language C++ [STR 00] – due to its proximity to language C – really democratized this concept. The language C is in fact renowned for programming onboard systems because of its proximity to the hardware. It has, therefore, been widely used and has earned a dominant position in the set of programming languages. The language C++, therefore, appears to be the "natural" successor of the language C in terms of reaching a high level of complexity in computer programs.

Similarly, countless design methods have appeared to model computer programs via objects: Booch [BOO 93], Classe-Relation [DES 94], HOOD [ESA 06], Merise [TAR 83], OMT [RUM 90], OOSE [JAC 92], etc. A collective effort for normalization, led by the *Object Management Group,* resulted in the *UML* in 1997. In 2000, this norm was accepted by the ISO and was updated in 2005 [ISO 01]. This work combined the three most fashionable methods of the mid-1990s: the Booch, OMT and OOSE methods.

Nevertheless, it is important to point out that UML is a norm that is still subject to certain adjustments. It has been in a constant state of development since its first publication to the present day.

The *UML* enables computer programs based on the concept of objects to be specified, designed and developed. It formalizes how different artifacts associated with software development are written: scope statements, system and interface requirements, how to split the system into subsystems, the formalization of processes and actors, the structural organization of data, etc.

The principal point of the UML is that it is a consensual (and normalized) language for exchanging information between different entities related to computer software development. However, the UML does not require nor specify any development method or any process to guide developers during the different phases of development.

The UML is designed to communicate through diagrams. These diagrams provide a graphic visualization of the ideas exchanged, facilitating the comprehension of the solution or the solutions proposed. Formalizing diagrams and notations reduces ambiguities, and thus incomprehension, as well as any resulting programming errors. What is more, the UML is independent of any programming language, any process and even any application domain; this helps make it universal.

1.2.2. *The Unified Modeling Language normalized diagrams*

In its normalized 2.0 version ISO/IEC 19501:2005, the UML defines 13 types of different diagrams that can either be used or not used according to the needs and requirements of software developers. To refresh readers' memories, a quick overview of some of the most commonly used diagrams will be given below. In fact, countless quality books have provided an exhaustive and detailed description of these diagrams (some of these works are quoted below and are listed in the bibliography).

The first type of diagram to be discussed is the most well-known and the most used: the class diagram (an example of this is given in Figure 1.1). This diagram is a collection of static modeling elements that describe the structure (or the architecture) of a computer program independently from temporal aspects and from the dynamic of the system. The basic element is the class (giving the diagram its name), which is associated with a digital entity and a name. For instance, a class named "Person" can be associated with a set of data describing a person; a class named "Oracle" can be associated with functions generating predictions from parameters, etc. The second basic element in class diagrams is relationships. A relationship associates two classes and can even contain certain properties (noun, cardinality, navigation between the classes, roles of each end of the relationship, etc.). Other elements complete class diagrams: dependency relationships, packages, etc.

Class diagrams are widely used in software development to structure programs, the data they process and associated databases. The design and analysis of class diagrams are the main focus of countless works related to the UML, to the extent that sometimes other types of diagram are eclipsed.

The second diagram is the object diagram. This diagram represents objects, i.e. concrete instances taken from classes at a specific time. This representation gives a more or less precise picture of the state of the object and its relationships with the other objects at a system's "photography" or "snapshot" moment. It helpfully completes a class diagram showing how the latter can be used. Therefore, the example of a class diagram (Figure 1.1) represents the (simplified) decomposition that was adopted to implement the Secure Next Generation (SNG) router described in Chapter 3. This example can be shown as the object diagrams presented in Figures 1.2 and 1.3, which show two very different configurations of the SNG router deriving from the same software development.

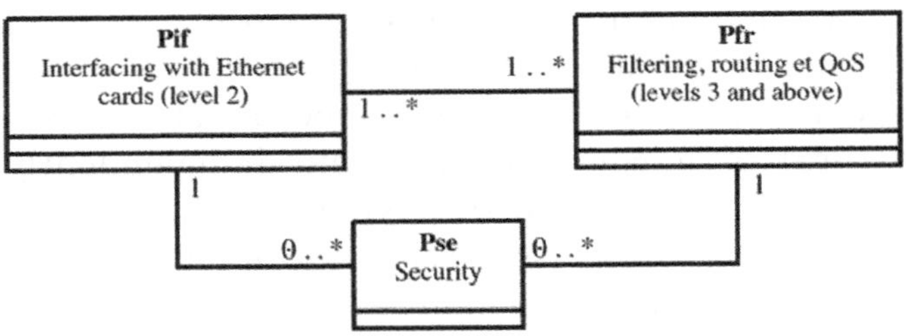

Figure 1.1. *Class diagram describing the software structure of the SNG router*

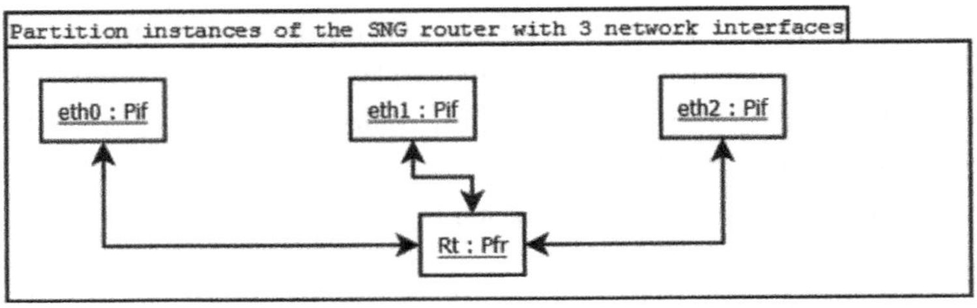

Figure 1.2. *Example of an object diagram deriving from the previous class diagram*

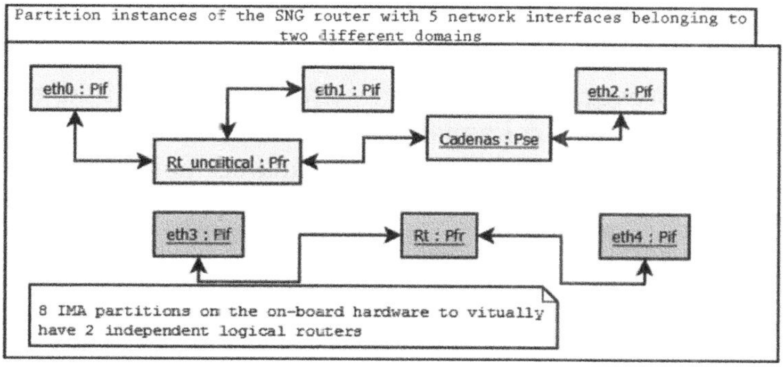

Figure 1.3. *Another example of an object diagram deriving from the previous class diagram*

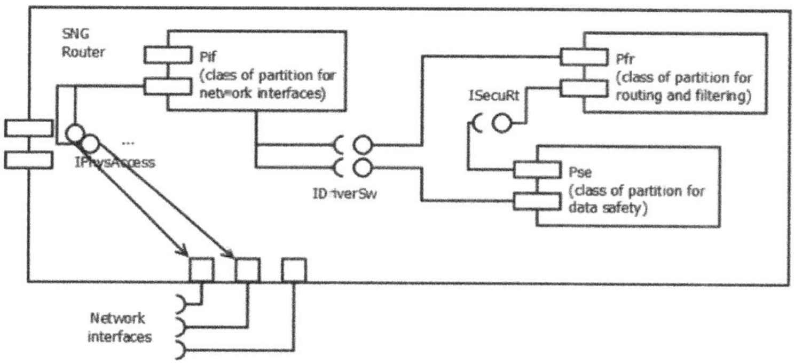

Figure 1.4. *Component diagram for the SNG router*

It would have been possible to use another type of diagram to represent the subsystems, the "components", that constitute the SNG router. The component diagram further explores the structure of a system by emphasizing the "black box" aspect of each subsystem and of the interfaces between the components. This gives the diagram presented in Figure 1.4. Though the UML offers tools that have no effect on the freedom of use, in practice the component diagram is used to obtain a general and undetailed view of the system; the class diagram meanwhile is used to detail each subsystem and each set of data.

All of these diagrams are capable of describing a system: from an elevated, abstract point of view down to the smallest details. However, the diagrams are atemporal and do not capture the behavioral aspects of the software system. In the UML, behavioral aspects can be tackled using behavioral diagrams.

The first diagram in this category is the activity diagram. Each activity in this diagram is a sequence of steps. The end of the execution of one activity initiates a transition that leads to the start of the next activity. The activity diagram can include conditions (conditional connection toward an activity or another according to the condition at the moment the transition is made), simultaneously initiate several activities or wait for their conclusion.

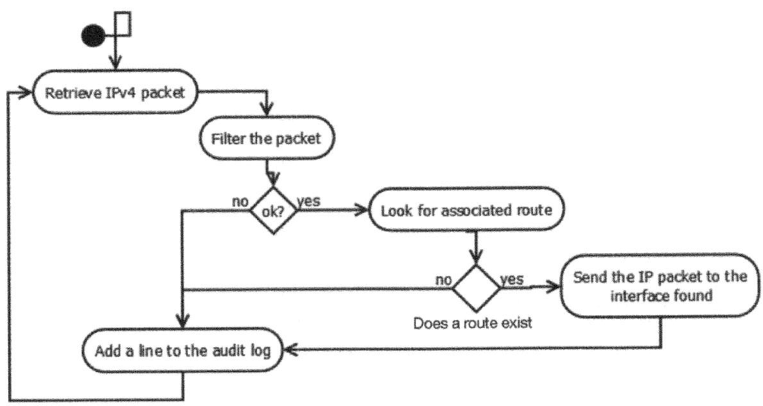

Figure 1.5. *Activity diagram for partition class Pfr4*

Each activity can be described in another activity diagram or using another type of diagram: the State Machine diagram. The latter is the direct descendent of diagrams that explain how finite-state automatons work. The diagram comprises the states and transitions (hence, its name) and describes how a system (objects and components) responds to interactions with its environment by modeling the different states of the system and the transitions between the states.

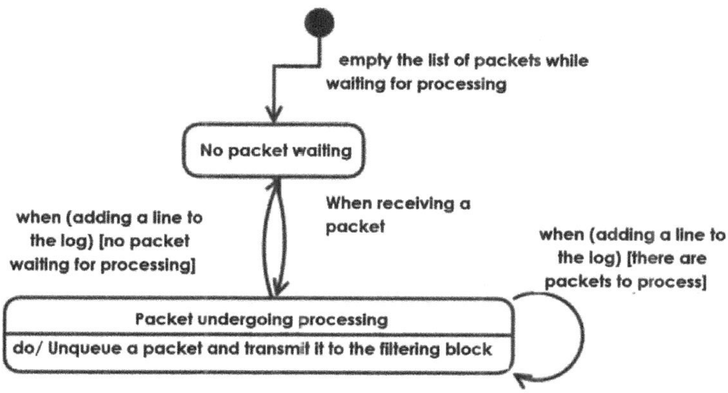

Figure 1.6. *Example of a State Machine diagram derived from Pfr4*

Nevertheless, though activity and State Machine diagrams can specify a summarized version of program behavior, acquiring an understanding of the program can require certain logical chains to be extracted that emphasize interactions between entities so that, for example, use cases can be illustrated. The most commonly used diagram for this task is the sequence diagram.

A sequence diagram is comprised of entities (objects or external actors, such as a human operator) that are arranged next to one another. The order in which they are arranged is not significant. Below each entity, there is a vertical line that represents the passing of time. Messages, shown as arrows, then travel from the vertical line of an entity to the vertical line of another entity to indicate that there is a communication between these entities. An example of this is shown in Figure 1.7.

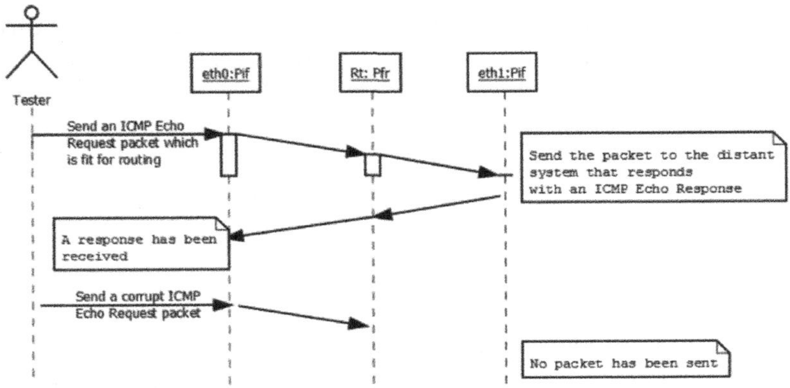

Figure 1.7. *Sequence diagram, used to test the SNG router*

Other types of diagram are specified by the UML norm. Composite structure diagrams show the relationships between components. In this diagram, the components are seen as white boxes. Deployment diagrams can show the physical and geographic arrangement of the entities comprising a complex system. Package diagrams show how entities and their interactions are clustered. Communication or collaboration diagrams are similar to sequence diagrams in terms of content, but they have different semantics: communication diagrams favor exchanging messages between objects, whereas sequence diagrams emphasize chronology. These diagrams, however, have not been used here so, if interested, the readers should consult the bibliography at the end of the book.

There remains one final type of diagram to be introduced. The use case diagram shows, in a graph, the user actor of a system or a computer program and the actions it can take. The relationships between the actors and actions are then translated into arcs that connect the actor nodes to the action nodes. This diagram is useful as it enables a client, a specialist in their profession and novice in software engineering, to express their requirements to a developer, a specialist in software engineering and novice in the professional domain of the client. Therefore, the use case diagrams are commonly used to draw up a scope statement so that the commissioner and contractor are able to communicate. Figure 1.8 shows an example of this.

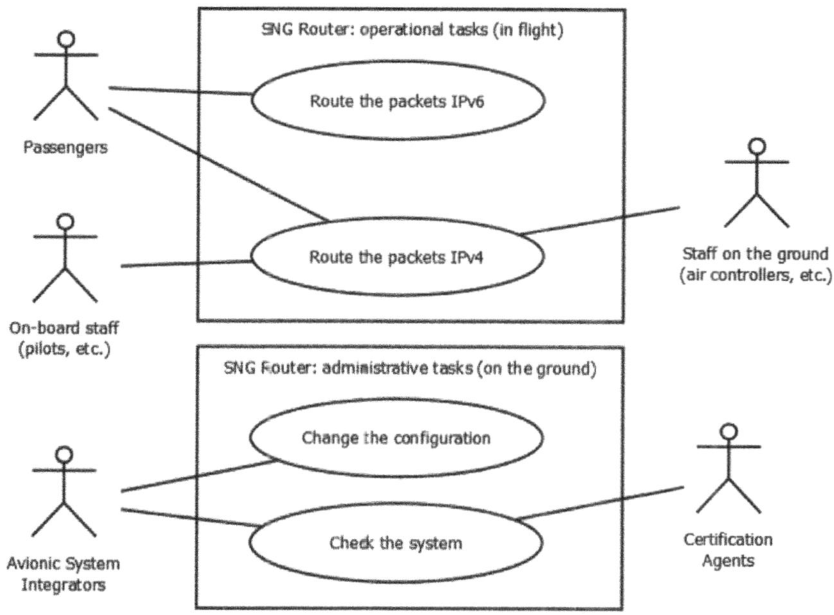

Figure 1.8. *Use case diagram for the SNG router*

We decided to discuss this diagram on its own as it is the only one, to our knowledge, for which no tool for the generation of software source code exists. In fact, all the other types of diagram have tools that can automatically generate code or offer a code skeleton or configuration files, etc.

To conclude this introduction to the UML, we would like to point out that there are many books and websites that develop and discuss in detail the different features of the UML. In addition to the official norm [ISO 01], we would like to point the reader in the direction of the following book by the founders of the UML [JAC 98].

1.2.3. *The advantages of model-based design*

The decision to fuse object-oriented design methods to form the *UML* is proof of the maturity of these approaches. Indeed, the concept of objects – tested since the dawn of computer science – has proven effective for managing complex polymorphic software development and is in a perpetual state of development. The classic functional approach (that consists of decomposing a system into functions and subfunctions) has proven to be harder to implement on a large scale and difficult to maintain over the mid and long-term. Therefore, the software industry turned to objects, which are more easily manipulated, maintained and reused over time.

Model-based design has, therefore, become a guarantee of the undeniable levels of effectiveness that are seen today. The UML acts as a support: it provides different views of the system via structural and behavioral diagrams that put the important aspects of the software or the solution to be produced into perspective. Normalized and with clearly specified semantics, the UML enables developers and the users and commissioners of the software to communicate with ease.

Object languages have had decades to become stable, and they are currently widely approved. The aeronautical world, however, remained cautious about object languages that were initially deemed to be too “young” and immature, and therefore potentially too dangerous to use. The qualification of the compilers[1], which is practically mandatory at the highest level of software certification, therefore prevented object approaches from entering the world of aeronautical software development. That said, in 2012 the DO-332 [DO 12c], entitled *Software Considerations in Airborne Systems and Equipment Certification*, was published. This document specifies the certification process for software using object-oriented languages and its application henceforth.

However, communication is not the only use of modeling languages such as the UML. Indeed, its strong semantics enables developers to use models and tools to generate and validate source code. In the 1970s, compilers grew in popularity and quickly became an essential tool in programmers' toolkits, next to linkers and assemblers. In a similar manner, code generators deriving from UML models or models in other languages are, in turn, growing in popularity and becoming essential.

We will call these code generators "transformers" to differentiate them from the "compilers" that are also generators: compilers generate binary code from the source code in programming languages, such as the language C; transformers generate source code in languages such as the language C from – often graphic – models, such as UML models. As such, the transformer → compiler → assembler → linker chain of tools enables the construction of computer software to be automated.

1.3. Formal model-based validation techniques

Models are able to describe entire software solutions with precision. All or part of the software code can, therefore, be

1 The qualification of a compiler is an optional procedure during the certification of an aeronautical software. The aim of the qualification is to reduce the number of steps involved in the verification of software artifacts.

generated in an intermediary language (the language C, C++, Ada, etc.) from these models. Initial models are also precise enough to verify and validate (or invalidate) certain software properties.

Indeed, at an early phase, models are able to verify whether the software or some of its functions could get stuck (no undesirable infinite loop), whether numeric calculations risk failure (the hardware limits of integer coding have not overflowed) and whether some states are specified although they are inaccessible (such states and their associated codes are said to be "dead").

The methods for verifying these properties in underlying software models can be classified into three categories and will be described below: *Model Checking*, formal theorem proving and code assertion methods.

Whichever method is used, the verification tool outputs three possible results: a verified-and-validated property, a false property (this result may potentially be accompanied with the counter-example that has invalidated the property) and a neither-validated-nor-invalidated property (when the verification takes too much time or cannot conclude due to insufficient data).

1.3.1. *Model Checking*

The first of these methods, *Model Checking*, consists of establishing the set of states through which the program can pass and then browsing this set to validate the property to be verified.

Let us, for example, consider the following algorithm (in the Pascal language) for the retrieval of an integer in a table of two integers between 1 and 3.

```
Function Search_integer
(N_to_to_search : 1..4;
Tab : array [1.. 2] of 1..3)
: boolean ;
Var i : integer;
    F : boolean;
Begin
 F := FALSE;
 If F Then
  i := 5;
 For i := 1 to 2 do
  If Tab[i] = N_to_to_search then
   F := TRUE;
 Search_integer := F;
End;
```

Algorithm 1.1. *Algorithm for the retrieval of an integer in a table*

In this simple example, the *Model Checking* tool will, in the first step, construct the set of states. At initialization, there are 36 possible states (four possible states for N_to_to_search x 3^2 for Tab, the states resulting from the initial values of i and F will not be counted in this example). Then the different executions of this function are made in such a way that each of these states is followed by the activation of another state (corresponding to the next instruction) and so on, until the end of the function, as represented in Figure 1.9.

In the second step, the Model Checking tool is, therefore, able to verify that:

– all the initial states lead to the final state and either return as TRUE or FALSE (the algorithm never "gets stuck" in an endless loop);

– F can take no other value than FALSE and TRUE;

– i only takes values 1 and 2, whichever succession of states the function passes through. There is, therefore, never an operation that results in the capacities offered by the *integer* type being surpassed;

– line 7 is dead code: indeed, branching line 6 never leads to line 7, regardless of the initial state of the function;

– the initial states with N_to_to_search = 4 always lead to the function result being FALSE.

Though simplistic, this function shows the limits of some *Model Checking* tools. Indeed, an exhaustive analysis of the set of possible states in this function would lead to a combinatorial explosion in cases where the table is larger in size and has "wider" bounds for the data contained in the integer Tab table (Algorithm 1.1). For instance, in a table of 16 cells that each contain a byte, there are 2^{130} possible initial states (still excluding the states of F and i), therefore, of the order of 10^{39} states.

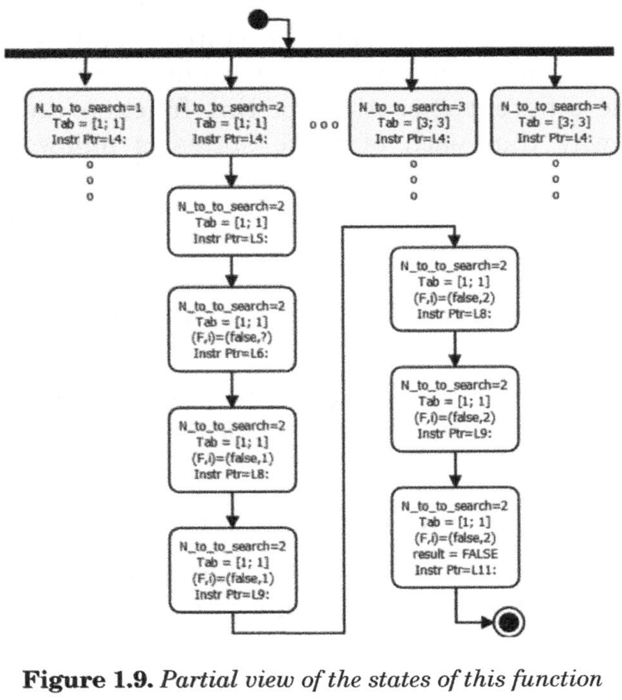

Figure 1.9. *Partial view of the states of this function*

However, it is possible for this function to group certain states together. Therefore, if N_to_to_search = 4, then the processing executed on a cell of the integer Tab table is identical for any value other than 4. For the branching of line 9, the cell Tab[i] = 1, = 2, = 3, = 5, = 6, etc. can, therefore, be grouped together.

In the same way, the initial state of F has no importance because at line 5 F is FALSE. Therefore, the *Model Checking* tool can use a specific value – "whichever" – for F while the state graph for this function is being constructed. The initial state is not duplicated because there is no need to process the initial states F=FALSE and F=TRUE differently. Grouping in this way enables the size of the state graph for function to be drastically reduced and the combinatorial explosion phenomenon to be compensated for.

A good *Model Checking* tool, therefore, maintains the most reduced space of states possible during construction so that it can verify the properties in a reasonable time. It is worth pointing out that what is understood to be a "reasonable" execution, as a human parameter, varies from one software project to another and in practice can range from seconds to months.

Model Checking tools use the space of states of a function. This space can be represented in a UML state machine diagram. Similarly, some *Model Checking* tools take a UML state machine diagram as input and directly validate properties on this diagram. Therefore, the state machine model, which is used by developers to design functions and automatically generate part of the software source code, also serves to validate this source code.

1.3.2. *Formal theorem proving*

In addition to *Model Checking* methods, the academic community has focused on another line of research that seeks to guarantee the properties of software modules. This line consists of extracting a theorem from the module and formally validating this theorem, hence its name "formal theorem proving".

It is, therefore, possible to extract the following theorem from the function of the previous example: $\models (i>0) \wedge (i<3)$. The function and theorem to be proved are given to the verification tool and the tool sends back its results (in this example "validated" because the variable i only concretely takes the values 1 and 2).

Formal theorem proving complements *Model Checking*: some properties that are unverifiable with one of the methods are sometimes trivial to verify with the other and vice versa.

Internally, formal proving algorithms are often complex algorithms that rewrite the property to show the equivalence of this property joined to the input data with the result.

In the same way that UML state machine diagrams are similar to *Model Checking*, it is possible to provide UML class diagrams annotated with *Object Constraint Language* (OCL) and possibly the source code of the property as input for the formal theorem proving tool. The tool then validates or invalidates the property. OCL provides constraint expressions on elements and entities in the diagram. For more information on and quality examples of this language, please consult Chapter 4.

Other formal theorem proving approaches consist of initially writing theorems, abstracts and then successively refining them so that after several iterations the source code is written. This source code is concrete and can be compiled and then executed on a real machine. The formal proving tool, therefore, guarantees the equivalence between the theorem, all successive refinements and the source code. These approaches use text languages such as Coq [BER 14] and B [ABR 96]. Following the publication of DO-333 [DO 12d] in 2012, which relates to the certification of software using formal methods, these formal methods can be used for development in aeronautics. Nevertheless, these approaches differ to those applied to the framework of the SNG router case study that will be presented later. They, therefore, fall beyond the scope of this book.

1.3.3. *Code assertion*

The third and final category of formal methods that guarantee the aeronautical software is working correctly is code assertion. While *Model Checking* and theorem proving both chiefly arise from academia, in the 1990s industry was calling on academics for solutions to validate code without the need to rewrite models nor to develop the same piece of software from scratch.

The chosen solution involves adding properties to be validated by directly annotating the software's source code. These properties are written in the source code comments so that the execution of the compiler remains unchanged. These properties adhere to a specific formalism recognized by the tool that formally validates the properties to be verified. This tool does not modify the code: validations are realized in a static manner without dynamically executing the code on the real system. In terms of terminology, this is the *static analysis* of the source code.

For commonly verified properties, preconditions, postconditions and invariance clauses are added to the function to be validated, in accordance with Hoare's 1969 logic model [HOA 69]. Preconditions are data that are provided to the verification tool to guide it in its evaluation and inform it of the state of the environment of the function. Likewise, invariance clauses indicate what the function should not modify. Finally, the tool formally verifies that the set of preconditions, invariance clauses and the code of the function to be tested systematically imply the set of postconditions. If this is not the case, the property is invalidated. The tool indicates this to the tester and shows him or her the reasons it failed.

The formalism that is typically used to constrain source code in language C is ACSL [MOY 08 and BAU 09], ANSI/ISO C *Specification Language*. This formalism can be verified using an automatic tool such as Frama-C [FRA 13]. Internally, this tool exploits a set of modules to realize the abstract interpretation of the code and then formal theorem proving and *Model Checking* (the last two have been presented in the previous sections).

Code assertion approaches have, therefore, enabled industry to become familiar with formal method concepts and to integrate tools, such as *Model Checking* and theorem proving codes, more serenely. This approach is an interface somewhere between pure computer science developed in an academic setting and application-based computer science arising from the immediate needs of industry.

1.3.4. *Applying for certification for complex systems*

Unlike object technologies, which were for a long time officially ignored by aeronautical certification institutions, formal methods were quickly addressed in international certification norms.

Indeed, in 1992 the publication of DO-178B [DO 92] explicitly authorized the use of formal methods. However, the concision of this authorization (just two paragraphs addressed the use of formal methods) raised countless questions about the methods to be implemented and the procedures for validating the use of formal methods for aeronautical software. Industry, therefore, favored other methods, such as exhaustive tests, which were judged more likely to convince certification agents of the innocuity of avionic software.

Consequently, until 2012, certification agents only had a few judicial examples of aeronautical applications validated by formal methods at their disposal. The committee that worked on DO-178C [DO 11], therefore, involved scientific experts in formal methods to expand this point in the new version of the standards. So in 2012, DO-333 [DO 12d] was published, an appendix to DO-178C that relates to the use of formal methods for the certification of aeronautical software.

The use of formal methods offers a myriad of opportunities, ranging from the validation of the software to each stage of its development, primarily the testing phase. Exhaustive tests are always limited by the performance of the computers testing the software and the complexity of the software. Indeed, the latter are becoming increasingly complex because computer systems are calling for ever increasing numbers of functions, there are new interactions between systems that were previously isolated from each other and the quantity and variety of the input data of software is growing.

Whether it is by *Model Checking*, *formal proving* or *code assertion*, formal methods give a formally (and often automatically) verifiable proof. This rapid and synthetic proof replaces "classic" software testing and thereby modifies the workload of testers who are transformed from developers of long tests into developers of short properties to be tested. DO-333 instructs development teams and certification bodies on which manual verification stages can be completed and even entirely replaced by formal automatic methods.

Concretely, the shift from manual tests to automatic tests by formal methods comes back to a shift from manual code to model-generated code. Manually written artifacts (whether they be operational code or tests) must be validated by reviews and/or tools that have themselves been manually validated. Automatically generated artifacts (whether they are code generated from a model or the result of a model-based property validation) are permitted on the condition that the generator has been previously evaluated, i.e. it has been manually reviewed and validated. This intellectual step initially comes with a significant cost, but this is outweighed by additional benefits: reuse and automating the generation of proofs that are provided to aeronautical certification agents.

An original method of software prototyping based on model-based design techniques will be presented in the remainder of this book. The method also uses powerful formal verification methods for the certification of the complex systems produced. A concrete example of an onboard aeronautical system will illustrate how these new techniques enable both the development and validation of the produced system to be accelerated, while sticking to acceptable design and development costs.

2

Methodology for Rapid Prototyping Avionic Software

This second chapter will first present the specificities and the restrictions of the avionic domain in terms of both its functionalities (security, safety and system virtualization) and standardization (the need for a complete certification of the final system). Taking these specificities into account, a methodology of rapid development has been produced and will be presented in section 2.3. This methodology will be first described in the abstract and a list of its advantages will be given. Second, examples of tools enabling the implementation of its different stages will be provided. Specifically, the organization of the target architecture for the avionic software to be produced, modeling and transforming models into software source code will be described. Nevertheless, this instantiation example of our methodology is not unique and to conclude other examples of tool chains that can be used to meet the specific needs of each software product will be given.

2.1. The specificities of the avionic domain

The improved performance of aeronautical systems currently enables us to envisage the use of new technologies in the context of onboard aeronautical systems as well as avionic networks – which to this day have remained closed to ensure their security and safety – being opened up to public networks such as the Internet. With these new technologies come new security needs. Meanwhile safety must be maintained.

We therefore decided to work on a Secure Next Generation (SNG) router. This router enables the critical streams (such as communication between the pilots and controllers) to be multiplexed (as illustrated in Figure I.1 in the introduction) with uncritical streams (*streaming* for passengers) in one channel of communication and guarantees the security and safety of communications.

2.1.1. *System virtualization: integrated modular avionics*

The first generations of avionic software systems were based on the direct relationships between systems: when a sensor transmitted information to two onboard computers, the data were duplicated across two independent channels of communication, each with its own receiver. The arrival of new technologies provided crews with new services and introduced new interactions.

Figure 2.1 depicts an *Integrated Modular Avionics* system (IMA, see [GAT 09]), which executes several independent software systems within the same hardware module. Entitled *Integrated Modular Avionics (IMA) Development Guidance and Certification Considerations*, the RTCA norm DO-297 [DO 05] of 8th November 2005 provides regulation on the design and the implementation of systems for IMA architectures in civil aeronautics. Prepared by the *Special Committee 200* (SC-200) group, this norm defines IMA as "a shared set of flexible, reusable and interoperable hardware and software resources that, when integrated, form a platform that provides services, designed and verified to a defined set of safety and performance requirements, [shared set designed] to host applications [on the hardware] performing aircraft functions". This norm explains and separates the roles of the different suppliers of IMA modules: application suppliers, suppliers of IMA platforms, system integrators and certification agents.

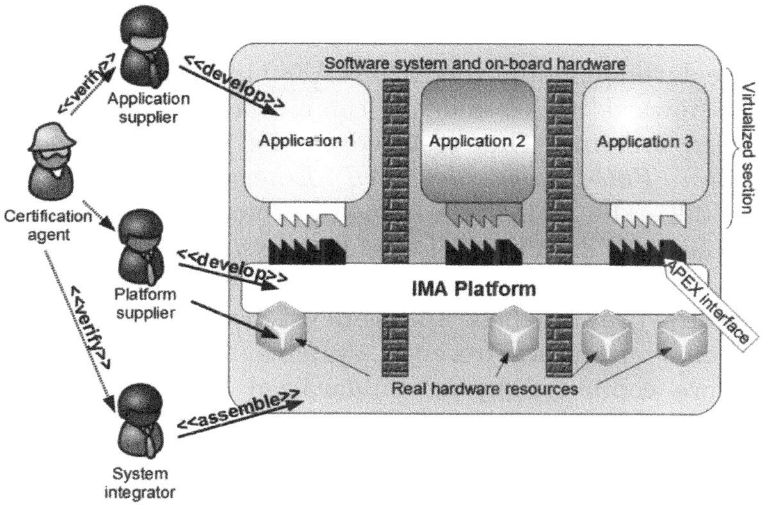

Figure 2.1. *The concept of IMA*

2.1.1.1. *ARINC 653 APEX interface: application executive*

In 1996 ARINC published the ARINC 653, a standard describing a programming and configuration interface that facilitates the independence of avionic applications with regards to hardware. The standard is called the ARINC 653 APEX (*Application Executive*). In 2003 an update was published that introduced IMA. The most recent publication *(ARINC653P1-3, ARINC653P2-2, ARINC653P3, RINC653P4),* in 2006, completes the previous version by clarifying certain points.

This concept therefore avoids duplicating physical connections: one bus destined for the hardware transports the multiplexed data for the different software applications. The growing number of avionic systems has, however, continued to require interconnectivity between systems, each of which potentially has different restrictions.

2.1.1.2. *AFDX bus: Avionics Full-Duplex switched ethernet*

The development of the Airbus A380 increased the need to move toward the scale of communication mechanisms between systems. This resulted in the introduction of the *Avionics Full-Duplex switched Ethernet* (AFDX, see [LAN 09]). The AFDX is a network protocol at the data link level (layer 2 "data link" of the *Open Systems Interconnection* [OSI] model), based on the reliable and redundant Ethernet protocol which enables the generic transmission of data between emitter and receiver systems. AFDX systems are therefore connected by a standardized Ethernet interface. Data are contained in the *Internet Protocol* (IP) packets, which are transmitted into the Ethernet frames. AFDX switches ensure the direction of the frames within the AFDX network.

However, the size of the topology of an AFDX network is limited by technology. The current closed network of the A380 contains a hundred or so terminal nodes which represent large amounts of work in terms of administration and certification. In addition, specific research is required to optimize the performance of this avionic network. Opening up this network to external data increases the number of potential nodes in the network. At a higher protocol level than that of the AFDX, the IP protocol (level 3 "network" of the OSI model) offers the opportunity to increase the size of virtual networks by maintaining acceptable-sized physical networks at the lower layers. The IP protocol is linked to routing functionalities: while "physical routing" (the switching of frames) is realized at level 2 by AFDX switches, "virtual" routing (packet routing) is possible at level 3 using a network component known as the "IP router". We have worked to design, realize and validate this type of router.

2.1.2. *MILS: divide and conquer, to rule over a secure world*

The segregation of the guests (see Figure 2.2 for example), brought about by a virtualization solution, has led to it being used to strengthen the security of sensitive systems.

The *Multiple [and] Independent Levels of Security [/Safety]* (MILS) architecture concept has come about gradually. Based on John Rushby's [RUS 81] concept of separation kernel, an MILS architecture guarantees a high level of security for the execution of programs in the same infrastructure.

Indeed, a founding principle of the engineering consists of decomposing a complex task into several more simple tasks. In computer science, this concerns decomposing/dividing a piece of software into modules. The evaluation of security is thus simplified because the evaluator does not need to evaluate a complete monolithic system but rather a set of smaller distinct modules and their couples.

Implementing a virtualization solution to realize the support of an MILS architecture is acceptable if it is proven to guarantee the four intrinsic properties of MILS:

– the solution is non-bypassable: i.e. a guest cannot communicate without passing through the safety controls imposed by the host system;

– the solution is evaluatable: the correct and valid operation of the virtualization system (and therefore the host) can be formally proven;

– the solution is always-invoked: communication is controlled not only for the first message but for every message exchanged;

– the solution is tamperproof: it prevents any modification that has not been explicitly authorized.

These properties are guaranteed by evaluating the security of the solution [OMA 05]. Evaluation is complex for "small" systems; and it is just as complex for minimalist systems (micro-systems), which are dedicated to system virtualization and their separation, known as "separation micro-kernels". These separation kernels ensure that temporal and spatial separation concepts between programs and the control of information stream are implemented.

The kernel ensures that each program and its virtual machine can use the real resources during the time that is assigned to them. A program cannot encroach upon another and "steal" its execution time. We therefore speak of the temporal separation between the two VMs.

The kernel also guarantees that a real-resource cannot simultaneously be assigned to two virtual machines. Address spaces for the memory and the input/output are shared during the configuration of the separation kernel: this is known as the partitioning of the address space. Then each time a resource is accessed, the kernel verifies that the virtual machine has authorized access and blocks the access attempt where necessary. The kernel therefore ensures the spatial separation between VMs.

Similarly, VMs can have channels dedicated to intercommunication. These channels are managed by the separation kernel (and not directly by the underlying electronics). The kernel then ensures the form of the communications: verification of the maximum length of the messages sent, access authorization, marking received messages, etc.

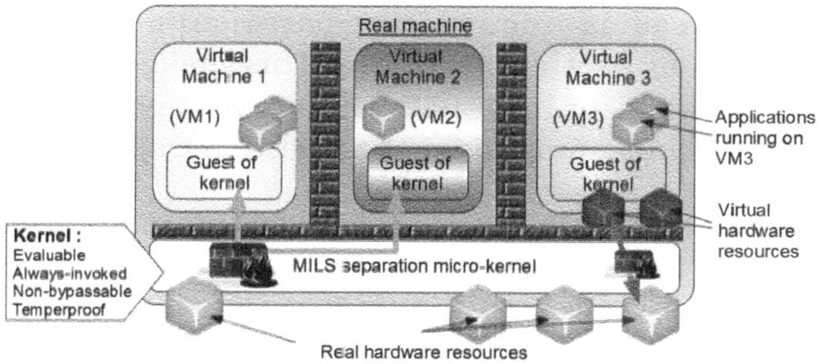

Figure 2.2. *Architecture of a MILS system*

As illustrated in Figure 2.2, each virtual machine in a MILS system is executed independently of its homologs. Seen from the host separation kernel, each VM has its own address subspace. The set of these subspaces is a partition (in the mathematical sense) of the overall address space of the host. Each VM has its own time slots, which are periodically repeated. The set of these time slots form a time partition. In the context of MILS architecture, the term "partition" therefore naturally designates the temporal and spatial resources associated to a virtual machine. During system execution, the separation kernel is, therefore, the partition scheduler and the non-bypassable controller of access to resources.

A MILS system must guarantee certain properties [UCH 09a, UCH 09b]:

– with regards to inter-partition information flows, only authorized sources can generate information and they are only delivered to explicit and authorized recipients;

– partition data are only accessible to the programs associated with the partition in question, they are isolated from the partition data of other partitions. Private data remain private and there is neither the possibility of infiltration (reading from another partition) nor exfiltration (writing data of another partition). This makes it necessary for each partition to have its own address space, where an address is only meaningful for the partition that it uses;

– the processor itself must not allow information to pass from one partition to another, whether it be via hardware caches or even via computing time measures. For example, there are attacks based on analyzing the number of processor cycles required to extract information from cryptographic keys, as shown in [KOC 96]. An MILS system must block, where possible, this class of attack;

– the malfunction of one partition must not impact the other partitions. It must be detected, contained and corrected.

Used in military and civil information systems, MILS systems are of increasing interest to the aeronautical world. Aeronautics is, however, already using virtualization, though with a slightly different objective: the mutualization of hardware, which, as discussed in the previous section, has high guarantees of safety. The current issue for aeronautics is that of being able to process these two aspects (security and safety) at the same time. This is the focus of section 2.1.3.

2.1.3. *Processing security and safety aspects simultaneously*

The development of systems installed onboard aircraft is strongly restricted, in terms of safety and security. In aeronautics, we talk of *safety* to describe the safety and even the security of operations in aeronautical systems, i.e. the intrinsic properties of systems that allow them to "resist" malfunctions. The term *security* designates the security of aeronautical systems, i.e. protection against intentional threats (pirate attacks, etc.).

The avionic domain has always been strongly restricted in terms of safety. The general malfunction of the device can in fact lead to its partial or total destruction, thereby endangering human life. Different aeronautical standards specify safety restrictions that must be respected if the device is to be authorized to fly. Critical onboard avionic software, automatic pilots for example, are therefore subjected to the RTCA[1] standard DO-178B [DO 92].

DO-178B specifies five assurance levels concerning the development of avionic software, called *Design Assurance Levels* or DAL. DAL-A is the most demanding level in terms of restrictions and is used for critical applications, the malfunction of which could be catastrophic, whereas DAL-E is a non-restrictive level reserved for applications that have no effect of the safety of the device. To be installed onboard and used in the device, a piece of software must have been previously evaluated according to the safety criteria; the verification known as "certification" is conducted by national institutions independent of developers. In France DGAC[2] is the certification institution for aeronautical systems.

Originally, aircraft manufacturers only paid thorough attention to safety. They believed that the complexity of onboard systems and the fact that avionic networks are closed was enough to guarantee their security. In the wake of the 9/11 attacks in 2001, security became a major worry for system design and new restrictions and practices were introduced to the world of aeronautics. Sensitive products must be validated from the perspective of security. The validation of security is called "evaluation". Security requirements can be grouped into requirement "packages"

1 RTCA: Radio Technical Commission for Aeronautics.

2 Directorate General for Civil Aviation (DGAC).

that enable them to be "evaluated" in the final product. In the CC ITSEC, *Common Criteria for Information Technology Security Evaluation* there are 7 packages that are uniquely constituted of evaluation requirements: they are known as "Evaluation Assurance Levels" or EAL. These requirement packages are numbered from 1 to 7. EAL-2 contains all of EAL-1 and adds extra requirements. This continues up to EAL-7, the highest assurance level of security that a product can receive in the *Common Criteria for Information Technology Security Evaluation*. This ISO norm, referenced as ISO CEI 15408 and often abbreviated to *Common Criteria* or "*CC*", enables a complete system to be evaluated with regards to security requirements. It is published in four parts [CC 09a, CC 09b, CC 09c, CC 09d].

However, despite the existence of these two standards (DAL *versus* EAL), certification (to guarantee the safety of a system) and evaluation (to guarantee its security) processes are currently processed in a parallel yet independent manner. Some similar verifications are therefore effectuated twice. Therefore, in addition to working on the problems concerning the certification and evaluation of our SNG router, we have researched the possibility of processing these two processes simultaneously through the software engineering methodology that will be presented later in this chapter (see section 2.3).

2.2. The certification of an avionic system

2.2.1. *The qualification of tools for certification*

The certification phase of avionic software is as costly in terms of time, personnel and money as the assurance level it requires is high. A Flight Management System (FMS) to automatically pilot a plane must be certified at DAL-A and can therefore take several years and hundreds of engineers to make and cost millions of dollars. In fact, DAL-A certification requires the device to complete 66 certification tasks, verified independently (i.e. by different and independent development teams). Conversely, the software controlling the *No Smoking* and *Fasten Seat Belt* signs only require DAL-D and its certification is therefore only subjected to 15 control tasks. The fact it has little effect on the safety of the device therefore makes it much less costly to develop.

During design there is a technique that reduces the cost of certification: it consists of automating certain control actions by using specific tools that guarantee certain safety properties. For example, using a DAL-A qualified compiler ensures that the binary code conforms to the source code. The "verification of conformity of the binary code to the source code" task, which is required for DAL-A certification, is therefore avoided. However, the tool (in this case a compiler) must itself be verified to ensure that it does not introduce any malfunctions: the verification of a software tool is called "qualification" and is realized in a similar manner to the certification of aeronautical software.

2.2.2. *Model-based development approaches in aeronautics*

As detailed in the previous chapter, traditional software engineering has been using model-based approaches (or *Model-Driven Development*, (MDD)) for a long time. The *Unified Modeling Language* (UML) [JAC 00] supports the most commonly used design methods in the industry (all domains) to realize the definition and implementation of classic software. More specific developments (the domain of onboard aeronautical or space software, for instance) cannot use this toolkit. Indeed, the main problem of design methods based on the use of the UML is the lack of consideration paid to certification aspects through the different tools that are used to produce the final software system. This is not necessarily a problem for applications for the general public, which can be developed using the UML (the design of web services [MIL 03] for example), or even for industrial applications that do not have strict restrictions [BAH 03]. Nevertheless, this aspect cannot be ignored with regards to critical applications: planes or satellites, for example. It is therefore necessary to consider the certification of the final product and integrate the tool chain used in the standardization process adhered to by this type of application, which has strict security and safety restrictions.

Research has been conducted into these aspects. [JAC 04], for example, used MDD approaches to facilitate the development of control commands for commercial airplanes. [HOR 10] applied MDD approaches for the development of IMA systems. [DUB 12] analyzed the consequences of using MDD for the production chain of a commercial airplane. The primary limit of these studies resides in the fact that their contributions always focus on one specific aspect of the development of a plane. In addition, [WIE 12] proposed an MDD approach for the complete development of onboard software for a satellite. These works demonstrate that it is possible to apply MDD approaches to the design of a complex complete system. Unfortunately, to our knowledge, no similar approach exists in the literature for the aeronautical domain. Our methodology intends to bridge this gap by proposing an approach as well as a set of tools that enable MDD techniques to be applied to the entire design of a complex avionic system.

In this chapter, therefore, we will consider specifically the certification process adhered to by the aeronautical industry when onboard software is designed to respond to a specific need. In this case, several standards must be considered. We are only interested in those relating to software engineering for onboard systems: DO 178 C [DO 11] and DO 331 [DO 12b].

2.2.2.1. *DO-178C: Software Considerations in Airborne Systems and Equipment Certification*

This document was published by the RTCA in 2012. It is the fourth version of a document that lays out the standard that must be followed for the certification of an airborne software system. This new version considers the most recent advances with regards to MDD for software engineering. In particular, for the first time in the aeronautical domain, it considers the possibility of validating these systems using formal verification methods that reduce the amount of unit testing required for the validation of the final product. This is a considerable improvement for aeronautical engineering as it significantly modifies the way in which aeronautical systems will be designed and produced in the future. It emphasizes MDD approaches which are capable of automatically generating source code from high-level models that define the functionalities and behaviors of the final system.

2.2.2.2. *DO-331: Model-based development and verification*

In addition to the standardization document DO-178C, an extra document, DO-331, suggests utility classes and methods that can be used to automatically generate source code and to validate high-level models at the beginning of the source code auto-generation process. DO-331 recommends the use of verification methods deriving from models. In particular, the RTCA recommends three types of mechanism to facilitate the verification process: Modeling Checking, formal proving and code assertion [GIG 12]. These verification techniques have been used for several years in other industrial domains and are now considered sufficiently mature to be used in complex environments such as aeronautical validation and certification processes.

2.2.3. *A sophisticated methodology for the development of complex onboard systems*

With regards to the development of the onboard software in our SNG router, certifying an existing router at the levels required by aircraft manufacturers would have proven to be very costly. Therefore, we have looked for solutions that would allow us to develop a new generation router and optimize costs. This led us to create a methodology for the development of aeronautical software and to apply it to our SNG router. The objectives of our methodology are to minimize the time spent on modeling, developing and validating the router and to minimize the effort required for certification (safety) and evaluation (security), while maximizing the security and safety levels the SNG router provides.

The router thus enables us to concretely test and refine our methodology while the methodology enables a fast and efficient development and validation of our SNG router.

2.3. Methodology of rapid development in seven stages

2.3.1. *Presentation of the different stages*

In order to apply this methodology, it is assumed that the software requirement phase has already been realized. As such, before applying the methodology, the software architecture has a set of requirements that specify the set of services the software must provide as well as the set of restrictions that it must respect. With regards to our SNG router, this set of requirements includes, notably, requirements concerning packet routing, their safety as well as the performance expected by the final system.

The methodology can be summarized by the successive application of the following seven stages (illustrated in Figure 2.3).

First, the software architect creates a partition for the set of requirements: he/she splits this set into different disjointed subsets so that the union of the subsets gives the initial set of requirements. This subset is then associated with what is called a "software partition". This is the *partitioning stage*. In the context of the SNG router, three software partitions have been extracted:

– routing and filtering requirements (Pfr4);

– data security requirements (Pse);

– requirements for interfacing the software with the hardware (Piface).

Second, the behavior of each "software partition" is represented using graphic models of dynamic systems and state machines. These models enable developers to test and validate the operation of the software and its subsections from the design phase. This is the *modeling stage*. We modeled the SNG router using Simulink [MAT 13a] and Stateflow [MAT 13b], two toolkits that work with Matlab [MAT 13c].

Third, the models are converted into source code using a qualified automatic source code generator. This is the *transformation stage*. We generated source code in the language C for the SNG router using Gene-Auto [TOO 08], a tool that transforms free source code and Open Source models.

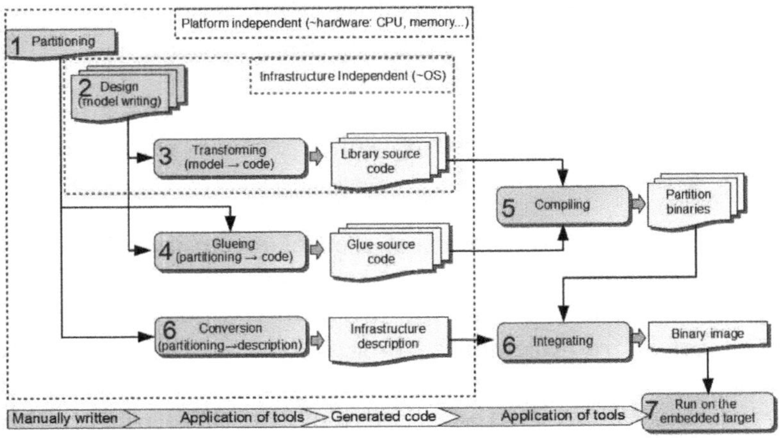

Figure 2.3. *Methodology of model-based development for complex onboard systems*

To work in a system, the generated code must be linked to an operating system (OS). To do this, another code, known as *glue code*, is generated. Glue code contains the links between the input/output of the model, the links for the software partition of the OS and scheduling information for the different partitions and the code entry point (for the previous stage). This is the *glueing stage*. In the onboard domain, some real-time certifiable and evaluatable systems exist, such as PikeOS [PIK 11], commercialized by Sysgo. We used this system to make our software modules work for the SNG router.

The fifth stage is the *compilation stage*: source codes generated by stages 3 (transformation) and 4 (glueing) are compiled and linked together in a binary code by partition. The use of a qualified compiler at this stage reduces the cost of certifying the final software product.

In the sixth stage, the software architect describes the overall structure of the software product by specifying for each software partition the number of instances that must be executed simultaneously, their configuration, their scheduling, etc. It groups this information, the kernel of the OS and the set of partition binary code (generated in the previous stage) into a "final image". This is the *integration stage*. This stage is often simplified by development tools provided by the OS, such as the Eclipse CODEO module provided with Sysgo PikeOS.

Finally, developers are able to download this final image on the onboard hardware. They can even execute it using an emulator in order to verify its operation and validate its performance. This is the *execution stage*. The qemu-x86 [QEM 11] emulator therefore enabled us to validate the correct operation of the router (as will be seen in the following section) with four domains communicating in IPv4 and IPv6.

2.3.2. *Advantages of the methodology*

This methodology reduces the load of the software development process in different ways.

High-level models enable designers to verify and correct the system's behavior "as early as possible": from the first phases of development. The risk of diverging from the acceptance test[3] (and the associated correction costs) and of avoiding malfunction and deviation is therefore reduced.

The functional decomposition into partition classes simplifies the design stage. It also isolates functionalities that have been incorrectly specified (for instance, where there are contradictions between the requirements) as well as those that have been imprecisely specified.

---

3 The "acceptance test" designates the final stage where the system is evaluated by the client.

This decomposition also facilitates the evaluation and certification phases: guaranteeing security and safety is easier for modular systems than for equivalent monolithic systems [OMA 05].

The concept of partition classes enables debugging partition classes to be designed to instrument the code. For instance, the development team can easily design a debugging class associated to an operational class, which will be certified, with the same inputs/outputs. Internally, the debugging class is linked to the operational class but it adds functionalities, such as displaying function calls, the values of inputs/outputs, etc. Given that the input/output interfaces are similar, the debugging class can be used in place of the operational class without there being an impact on the architecture and other partition instances.

Similarly, partition classes for security and safety can be inserted in the architecture to control the coherence of input/output, to verify the behavior of other partition classes, to correct the vulnerabilities that were not detected during the elaboration of the previous versions of the product, etc. The adaptability of these classes increases the security and safety assurances of the product. The last two points are illustrated in Figure 2.4.

The methodology can be automated from stage 3 (the transformation of code) to stage 6 (integration) to produce the final binary image. This automatization, in addition to drastically reducing development time, having perfect reproducibility and a determinism required by the certification processes, can accompany the qualification of the tools used at different stages. In this case, the tasks required for the certification of the system are consequently reduced and even eliminated. An example of reduced certification costs is given later in Table 2.1 (see section 2.3.3).

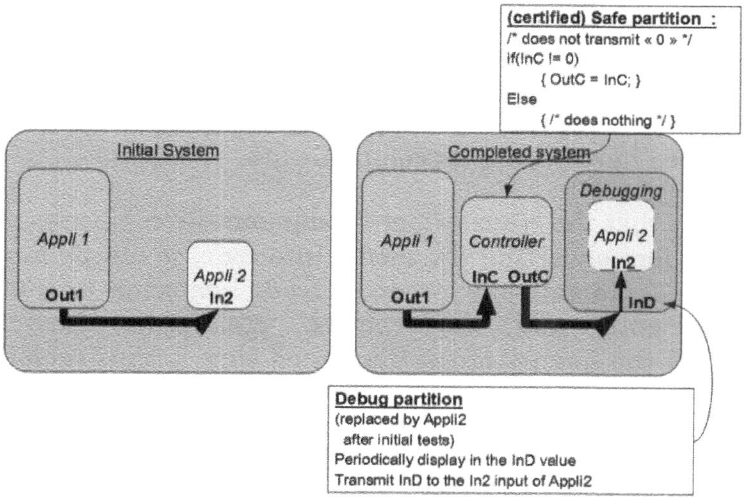

Figure 2.4. *Addition of security and burn-in partitions*

Automatically generating most of the source code at stage 3 (transformation) reduces the occurrence of malfunctions linked to repetitive tasks and bugs relating to copying/pasting. For instance, with the methodology, programmers no longer need to manually manage low-level tasks, such as controlling the access to shared memories or validating calculations, because together the semantics of high-level models and the different stages of transforming the model into binary ensure these repetitive tasks.

For example, a division by zero leads the processor to a potentially fatal fault for the execution of the software; it is therefore imperative that it should be avoided. The transformer can generate a source code in which each division is systematically completed by a non-nullity verification of the divisor: this pre-conditional verification avoids the breakdown.

Moreover, introducing hidden channels is even harder because an OS that systematically verifies all the inputs and outputs of the partition instances is used. Installing a hidden channel therefore requires modifying both the configuration of the OS and the two applications (or application and associated hardware) that communicate.

Given that the transformer fundamentally comprises a set of rules for converting the model into code, it can be proven that the automatically produced code conforms to quality requirements and development standards. Manually verifying the quality of the generated code is thus no longer useful.

However, if the transformer makes a coding error, the valid model is transformed into an invalid source code. This error is duplicated a large number of times, thereby invalidating the generated source code in its entirety. Transformers must therefore pass a tool qualification phase, which guarantees high levels of safety for the generated code.

The library code is independent of the virtualization infrastructure used and of the underlying hardware. This abstraction thus facilitates the task for the designer who does not require any specific skills in these domains and can instead specialize in modeling and in the connected algorithms.

The independence of the library code from the infrastructure has the undeniable advantage of portability: one library code can be reused on different OSs without requiring modification; only the glue code, a small part of the compiler source code, must be rewritten or regenerated. By way of example let us consider the partition class in the SNG router that ensures routing: the glue code is 350 lines of code (including comments and empty lines), that is, 1.25% of the total lines of code for this partition class. It is therefore possible to compare the performance of the code on different virtualization infrastructures quickly and with little effort.

This independence also facilitates the reuse of the code. The partition class can be imported from the original project to another project during the development of a new version or even a new system. It ensures the other functions and shares the subfunction, ensured by the reused partition class, with the original system.

The source codes (library code and glue code) are independent of the hardware: the compilation and integration stages will produce binary codes dependent on the hardware. The independence of the hardware platform has more advantages in terms of portability and reuse. So, from partitioning and models designed for the SNG router for onboard target hardware (bound for certification), we easily managed to reconstruct a derived version that is run on an office computer with a Intel® 64 bits processor. The derived version was not intended to be installed onboard but to be used as a prototype to demonstrate the functionalities of the router.

Modeling can be partially completed by manually written source code. This stage can be completely replaced by manually rewriting the library source code. In the event of the latter, writing the code also replaces the next stage, that of transformation. This freedom (for software developers) cancels out certain benefits with regards to certification, but it does allow for the reuse of source codes that have been programmed beforehand (also called "Previously Developed

Source code" in DO178C). This could be indispensable if the virtualization infrastructure does not provide adequate abstraction to be able to access certain specialized hardware resources. In this event, it could be preferable to write the peripheral drivers directly by hand in low and medium level (in the sense of "close to the hardware") languages, such as the assembler and the language C. We recommend, however, that the development project be architectured in such a way that these manually coded sections are isolated into dedicated partition classes. In this way, changes to the hardware of future versions will only impact these partition classes.

Using a virtualization infrastructure that is secure and safe and compatible with MILS [RUS 81] and IMA guarantees the segregation between the different partition instances and controls all interconnections. The number of vulnerability vectors in the system is thus reduced.

Moreover, strictly controlling the communications between the partition instances is associated with a small number of communication types. For instance, in IMA systems, thousands of communication channels are possible, though only two categories of channel exist: *message queues* and *sample queues*. The former is a finite queue of messages (which are dated and of a limited length) where new messages are added to the queue. The latter is a memory stamp that can host just one message at a time and where the new message replaces the old one. This facilitates the replacement of one whole software partition instance with a software solution with no impact on the other instances.

Finally, partition instances are intended to be executed simultaneously. The functional decomposition of the first stage directs the software architect toward designing a modular and parallelized system. Parallelizing the execution of models associated with functions opens the door to relatively unexplored perspectives in aeronautics. Although systems are currently generally based on monoprocessor and monocore hardware platforms, parallelization could fully exploit multicore processors and even massively multicore processors (for more information on this see the applicability of GPGPU[4] in avionics [GAL 11, AGR 11]) to enable a jump in the order of magnitude of the performance of future systems that will be designed by following this methodology. As such, our methodology is preparing for the evolution toward a new generation of platforms, that are yet to be created, on which each partition instance would have its own associated library.

This section has explored the general principles as well as the advantages of the methodology independently from the software tools used to implement it. We will now focus on the benefits of this methodology on a concrete example.

2.3.3. *Example of the benefits of using our model-based methodology*

With regards to certification, in compliance with the DO-178B norm [DO 92], there are advantages to using DAL-A qualified tools: 16 out of the 66 tasks required for DAL-A software certification are reduced and even eliminated when the transformer and compiler are qualified at the same level (see Table 2.1). Similarly for the evaluation of the final product, using a previously evaluated OS, such as Sysgo PikeOS, facilitates the evaluation of the final software product. Not only do certain parts of the evaluation not need to be re-evaluated, but the fact that the OS controls

4 GPGPU: General-Purpose computing on Graphics Processing Units.

communications between partitions is an undeniable advantage with regards to guaranteeing software security.

Task	Description
A.2.6	The source code is developed
A.2.7	The Executable Object Code is produced
A.5.1	The source code complies with low-level requirements
A.5.2	The source code complies with software architecture
A.5.3	The source code is verifiable
A.5.4	The source code conforms to standards
A.5.5	The source code is traceable to low-level requirements
A.5.6	The source code is accurate and consistent
A.6.3	Executable Object Code complies with low-level requirements
A.6.4	Executable Object Code is robust with low-level requirements
A.7.1	Test procedures are correct
A.7.2	Test results are correct, and discrepancies are explained
A.7.4	Test coverage of low-level requirements is achieved
A.7.5-7	Test coverage of software structure is achieved

Table 2.1. *Advantages of qualification for DO-178B certification*

2.3.4. *Instantiation of the methodology with different tools*

2.3.4.1. *Summary of compatible tool chains*

We applied the methodology to the SNG router using existing tool chains (see Figure 2.5 for details) for modeling (Simulink, Stateflow), transformation (Gene-Auto) and an OS (Sysgo PikeOS). Nevertheless, the methodology was created in such a way as to ensure that the development team has great freedom with regards to the tools used.

To illustrate the tool chains that are compatible with our approach, we would like to clarify that the methodology is not limited to the paradigms of imperative programming. Figure 2.6 summarizes the different tools set out in this section. The list is not restrictive as the methodology was designed to be as generic as possible.

It is important to state that no one column contains only one possibility; there are always several possible choices. In fact, the certification authorities demand that the most sensitive systems be made redundant and that the redundant version be dissimilar, i.e. developed by different teams using different programming languages and different tools. This is known as diversification.

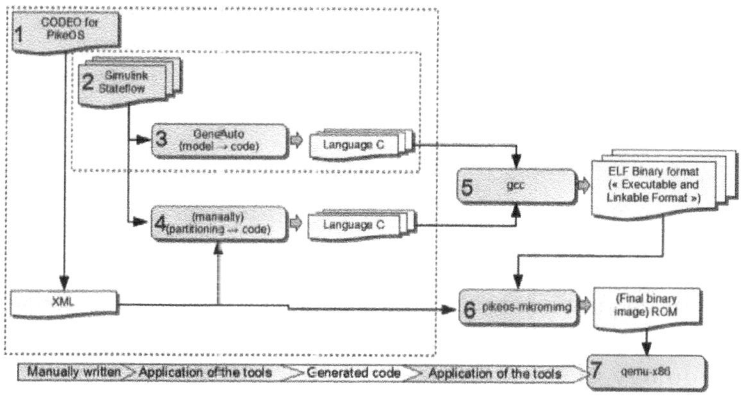

Figure 2.5. *Tool chains we used for the SNG router*

We can therefore envisage using functional tool chains with source code written in the CaML language [HUE 85], transformed into binary using an ocaml compiler [VOU 96] whose operation has been validated using formal methods [LER 06] or even concurrent OSs, such as the real-time system VxWorks [VXW 11]. We can also imagine using the method B [ABR 96] to model partition classes with the ComenC tool [CLE 13] for the generation of the language C. These aspects will not be developed in more detail in this book: the imperative tool chain presented in Figure 2.5 is perfectly adequate and is adapted to our example of the SNG router.

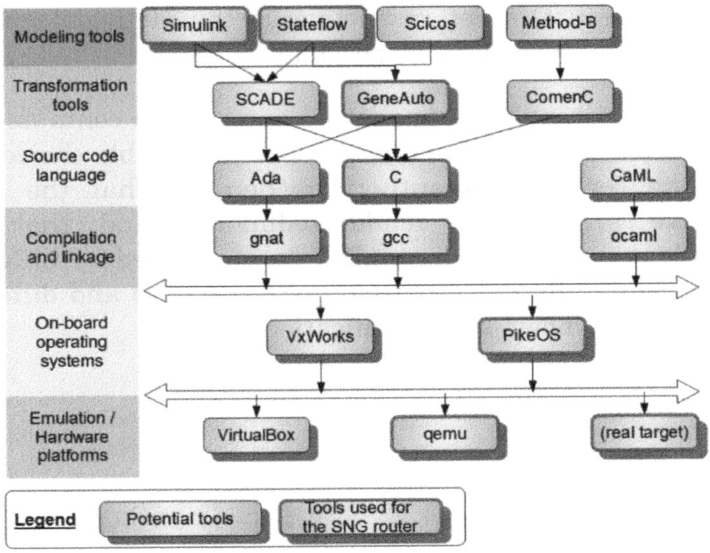

Figure 2.6. *Some tools that can be applied to our methodology*

We have just seen that one or several software tools can be associated with each of the seven stages in the methodology. The sequence of tools is thus called the "tool chain" as the methodology uses the outputs of the first tools as input data for the tools that follow. The next part of this section will illustrate the way in which the main tools have been used to realize the SNG router. Details regarding the design of the router will be dealt with in Chapter 3. The following sections have but one objective: to illustrate how the different tools (presented previously) can be used in the context of a specific industrial application.

2.3.4.2. *Organization of the architecture of the software solution*

The first stage in the methodology consists of partitioning the router functionalities. In this manual stage we used nothing other than a text processor (LibreOffice, see [LIB 13]) to partition the requirements into modules which will later correspond to the partition classes. The free software Dia [DIA 13] enabled us to quickly realize a class diagram that summarizes the architecture, i.e. in this case the set of partition classes and their interactions. This diagram is shown in Figure 2.7. The functional requirements were grouped into three partition classes: a "Pfr" class (Partition for filtering and routing) to route and filter the packets; a "Pse" class (Partition for Security) manages the security problems relating to streams (confidentiality, integrity and authentication); and a "Pif" class (Partition for interfacing) manages interactions with the hardware.

The three partition classes shown in Figure 2.7 can realize a router for simple topologies, such as topology 1 in Figure 2.8 where the architecture for the SNG router interconnects three domains.

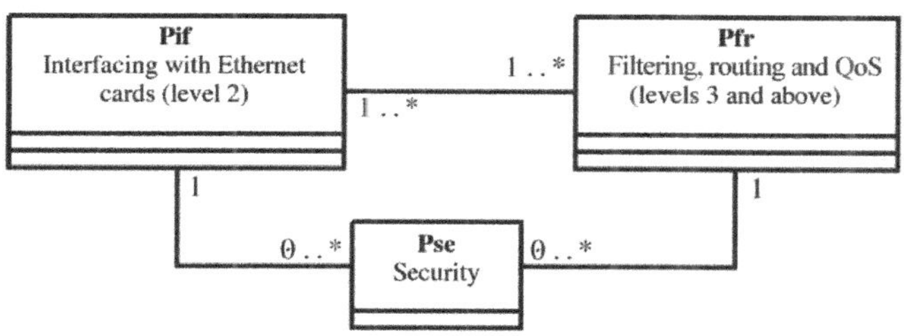

Figure 2.7. *Diagrams of partition classes for the SNG router*

It is possible to develop a second system by modifying the previous architecture without needing to repeat the modeling stage for the partition classes. Figure 2.9 shows a second network topology with five domains which are connected to two distinct and independent groups. The software for the SNG router is thus architectured again by adding new instances of existing partition classes (which are thus already certified and evaluated) and new connections between instances. Reuse reduces the work required for the validation of the new system.

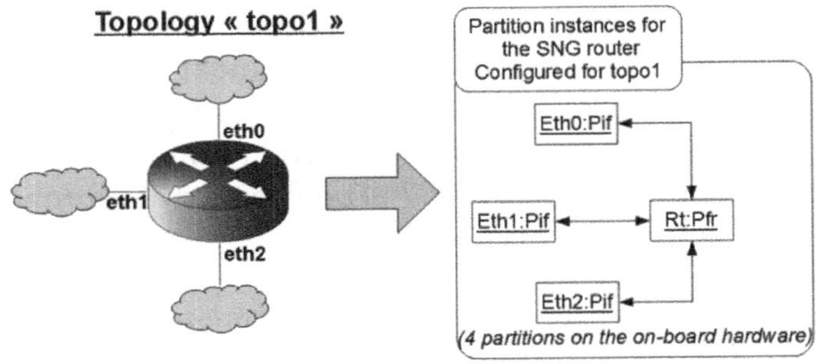

Figure 2.8. *Example of a topology where three domains are interconnected by the SNG router*

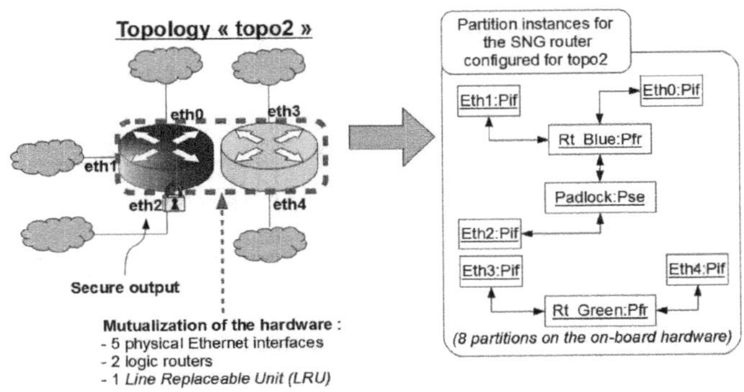

Figure 2.9. *More complex topology for the implementation of the SNG router*

A more complex project may find a software solution that is adapted to managing and following requirements to be advantageous in an application where countless professions intervene simultaneously. Rational DOORS is therefore an alternative we initially considered: it was not selected as we wanted the development of our router to be "simple" and light.

2.3.4.3. *Design: modeling with Simulink and Stateflow*

Once the partition classes have been defined, with the desired functions and their interactions, the second stage of modeling – using designers – can start. For the realization of a model for each partition class we used Simulink [MAT 13b] and Stateflow [MAT 13c], two extensions of Matlab [MAT 13a]. These software products modeled the module schema and the state machine diagrams respectively (examples are shown in Figure 2.10). They also simulated program operation and performed formal proving on the models to guarantee, for example, the absence of anything blocking the execution or infinite loops.

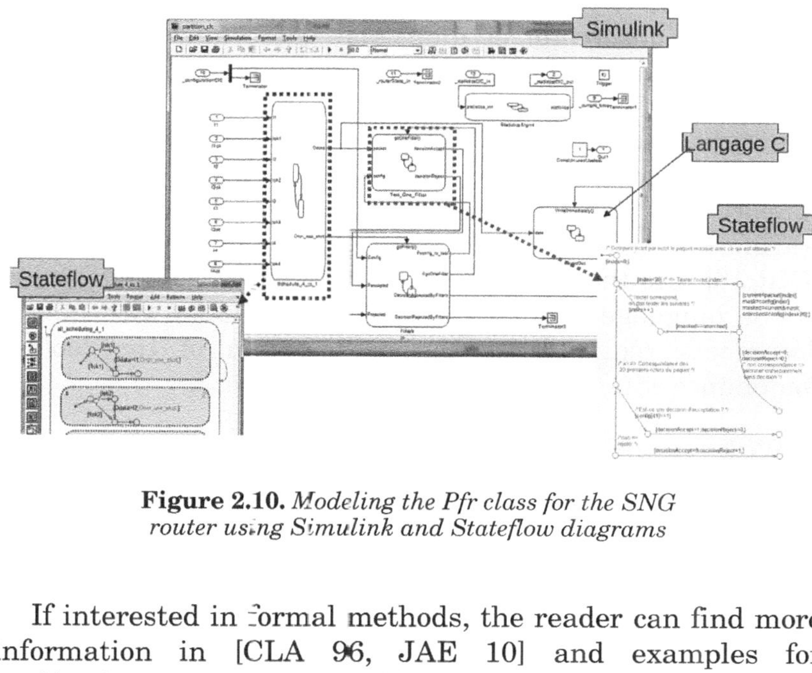

Figure 2.10. *Modeling the Pfr class for the SNG router using Simulink and Stateflow diagrams*

If interested in formal methods, the reader can find more information in [CLA 96, JAE 10] and examples for applications in avionics in [GHA 10, JOY 10, ADA 05] or [RAY 01]. With regards to MDD and its applications in aeronautics, we recommend [BUS 04, HOR 10, BHA 05, SUB 03, POL 11] or [DA 11].

It is important to point out that the sets of functionalities for the Pfr and Pse classes are disjointed; thus the classes can be modeled at the same time by different designers. This is the same for the Pif class, which, however, is manually written directly in the intermediate input language of stage 5 as it needs direct access to the hardware.

2.3.4.4. *Transforming into the language C with Gene-Auto*

The third stage consists of transforming models in high-level languages in source code into a language that is closer to the binary format the onboard target executes. To realize this transformation, we used the software from the Gene-Auto project, a free Open Source project managed by a European consortium of industrial users and suppliers of services and academic partners (for more information consult [JOB 11, RUG 10, RUG 08, TOO 08, IZE 08, TOO 10, IB 10]). We could also have used its competitor SCADE [SCA 13], commercialized by Esterel Technologie, but we preferred to support the free and Open-Source solution. Gene-Auto transforms the Simulink, Stateflow and Scicos models [INR 13] into the source code in the language C [RIT 72] or in the language Ada [ICH 83]. It is available with all the documents required for the qualification of the tool in compliance with the DO-178B [DO 92] norm.

With regards to the SNG router, Simulink models (containing the Stateflow models) for Pfr and Pse partition classes were converted into 56 files with a total of 10,500 lines of source code in the language C. Figure 2.11 shows an extract of the text output of this tool applied to the Pfr partition class.

The source code generated at this stage is a set of the functions in the language C, which is intended to be the library of the code executed in the partition. It is independent of the onboard hardware platform and even of the OS.

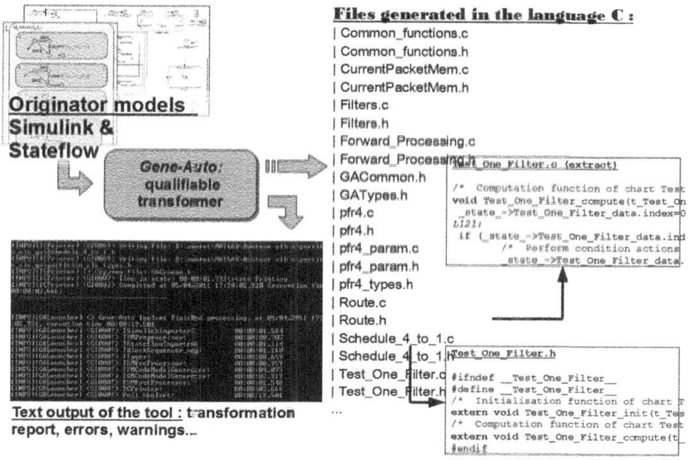

Figure 2.11. *Example of the automatic transformation of high-level models into source code using Gene-Auto*

2.3.4.5. *Execution with a strengthened onboard operating system*

2.3.4.5.1. Glueing and compiling code

The source code generated at the previous stage must be linked to the inputs–outputs of the system. Although some inputs–outputs are very dependent on the onboard hardware system, the majority of them are communications between partitions. The OSs that we use are compatible with the ARINC 653 APEX programming interface (see section 2.1.1.1 in this chapter). As a standard, this interface links the partition inputs with the model outputs in a deterministic way, regardless of the brand and the version of the OS. It does the same for the outputs.

The glueing stage has a secondary objective: to provide the OS with a unique entry point for the partition to be executed. The Gene-Auto transformator creates one function per Simulink bloc. What is more, the execution of these functions corresponds to the execution of one simulation step in Simulink: it thus stops at the end of the first iteration. However, the desired execution should infinitely start a new iteration. The algorithm located in the glueing stage is thus as follows:

```
/* AUTOGENERATED GLUE CODE FOR PARTITION PFR4 */
/* Do not modify, modifications may be lost...*/
/* GLUE MAKER INPUTS: Pfr4_on_PikeOS.xml       */

/*****  PikeOS Specific part *****/
#include <vm.h>
#ifdef PIKEOSDEBUG
#include <vm_debug.h>
#endif
VM_DECLARE_STACK(0x4000)

/*****  Model specific part *****/
#include "pfr4.h"

/*****  Glue code part for Pfr4_on_PikeOS *****/
#include "pfr4_glue_io.h"
#define FOREVER while(1)

/* All partition inputs and outputs */
t_pfr4_io io;
/* State of this partition   */
t_pfr4_state st;

extern void _p4_entry(void) {
  vm_init();             /* Init PikeOS PSSW */
  pfr4_init(&st);        /* Init Pfr4 model  */

  FOREVER {              /*  <Regular usage> */
    read_all_pfr4_partition_inputs(&io);
    pfr4_compute(&io,&st);
    write_all_pfr4_partition_outputs(&io);
  }
}
```

Algorithm 2.1. *Algorithm for the glue code*

The OS is told that this algorithm is the entry point of the execution for the partition. So, at the start-up of the partition, the algorithm is invoked and executed until the partition stops, resulting from either the extinction of the system or the beginning of a breakdown. OSs often need to use a specific name to identify the entry point; this identifier depends on the model and the version of the operation system, making the glue code dependent on it.

At the end of this stage, the glue source code is written and, as is illustrated in Figure 2.12, links the partition class model with the OS. The link is for the input and output of data and for the call of the adequate function of the model (for the jump of the Pointer Instruction from the OS to the application in user-land).

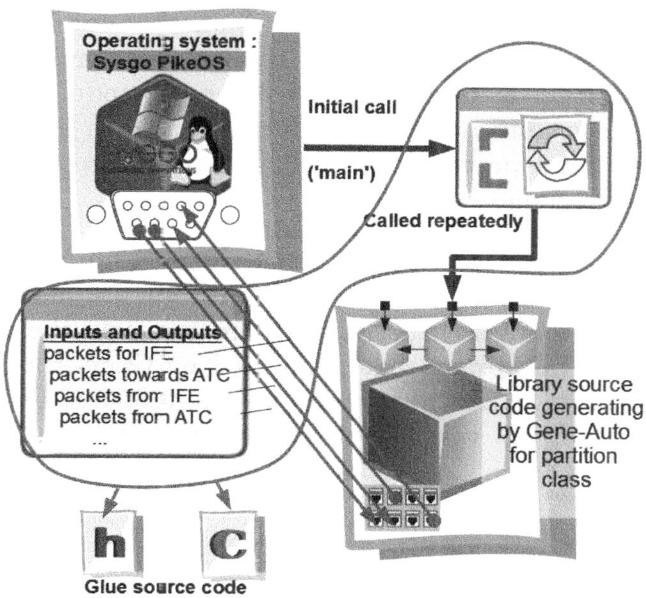

Figure 2.12. *Glue code linking the operating system to the partition class model*

When the library and glue codes are written, they are independent of any processor model associated with the hardware platform and of any address space partitioning. The compiler and the linker are the software tasked with establishing the adequate binary code, which corresponds to the source code, by using the instructions available in the instruction set (provided by the hardware platform's processor) and by partitioning the data and the binary code to the adequate memory addresses.

For the fifth stage of the methodology applied to the SNG router, we used a modified version of GCC [GCC 13] that had been fixed at an old tried-and-tested version. It would have been possible to use a more recent and optimized version, however this could have caused difficulties with regards to potentially qualifying the tool. We also tested the compiler gnat [ADA 13], created by AdaCore, with source code in Ada generated by an experimental version of Gene-Auto. This proved to be promising though incomplete: the transformation of certain types of model modules towards the source code in Ada was not implemented in this version. We were thus prevented from validating our experiments with the language Ada.

At the end of this stage we had a binary file by partition class that could run autonomously in a virtual machine. As illustrated in Figure 2.13, this stage converts the files *.c and *.h in the library code generated by Gene-Auto and in the manually written glue code into a unique and executable binary file that will be used for the instances of the Pfr partition class.

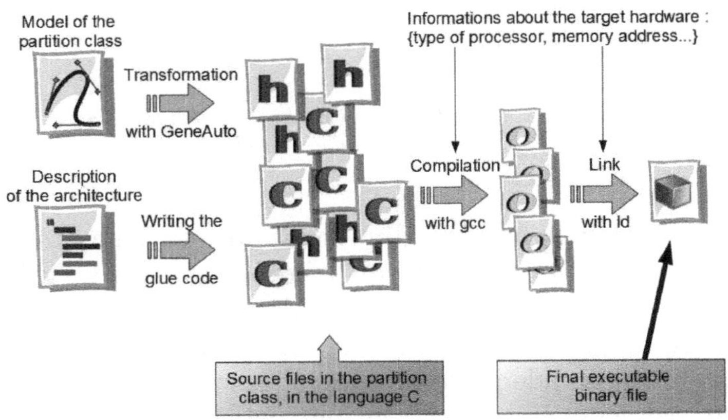

Figure 2.13. *Compilation and linkage of source figures of the Pfr partition class*

2.3.4.5.2. Integration with Sysgo PikeOS

The sixth stage of the methodology consists of integrating these binaries with an OS. For this stage, the choice of the OS has an impact on the properties of the final software:

– an OS that ensures virtualization (see sections 2.1.1 and 2.1.2) can integrate each instance of partition into a virtual machine;

– an MILS compatible OS (see section 2.1.2) has advantages in terms of the security of the execution of partitions as well as for the evaluation of the final system;

– an IMA compatible OS (see section 2.1.1) guarantees the safety of the execution of partitions and simplifies the certification of the final system;

– onboard OSs are provided in the form of modules that are either kept or not during the system elaboration phase. What is needed is therefore embedded;

– commercial OSs are sold with toolkits that are ready to be used (documentation, compilers, examples and tutorials, technical support, configuration help tools, etc.), which are undeniably useful;

– some OSs can provide real-time guarantees: these are necessary for some onboard systems, but in the end they are not much use for our SNG router.

Having researched different offers, we decided to opt for the PikeOS [SYS 10] system, created by Sysgo, as it provides the virtualization service, is MILS and IMA compatible, is sold to be qualified in compliance with the DO-178B and evaluated at EAL-6 (see section 2.1.3. for details), its micro-kernel has a small memory footprint (150 kB of RAM + 150 kB of ROM) and it is sold ready-to-use.

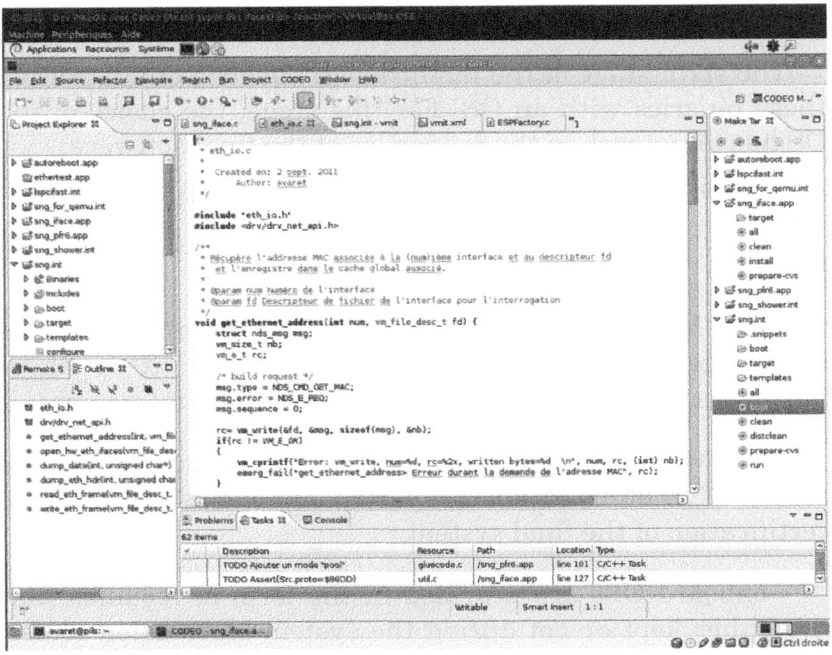

Figure 2.14. *Eclipse Rapid Application Development, extended with CODEO*

The VxWorks [WIN 09] system, sold by PikeOS's competitor WindRiver, has the same advantages and a better reputation; we did not choose this system because it was more expensive.

As mentioned in Figure 2.14, the Eclipse framework provides the embedded system engineers with different tools that improve their experience of system conception. Indeed, several wizards are avalaible to setup the system architecture and also to debug and test the different behaviors of their final system.

The toolkit provided with Sysgo PikeOS contains a Rapid Application Development (RAD), based on the Eclipse [ECL 13] application, extended with the Sysgo CODEO

[SYS 12] module. As shown in Figure 2.14, this plug-in facilitates the development of onboard systems by aiding design, by providing wizards to configure the architecture and by providing debugging and investigation mechanisms.

2.4. Chapter summary

This chapter focused on the development process that is applied to a complex onboard software application. Security and safety restrictions are critical in this regard. We therefore chose an OS that is compatible with MILS and IMA, two notions that were discussed alongside virtualization. These restrictions also pushed us toward using modeling tools, so that the system can be verified at the earliest stage possible, and toward automatic code generators, to avoid translation mistakes from a high-level language to languages that are closer to our target onboard hardware. These tools also enabled us to accelerate the development phase and to rapidly produce a functional prototype for the tests and the validation of the critical onboard software (SNG router). The latter will be presented in Chapter 3.

We chose to present the methodology, which is abstract and generic in nature, separately from its instantiation with the tools. The general principles of the methodology are in fact independent of the tools used and there are several candidate tools for each stage to be realized. The tools used for the SNG router were presented: Simulink and Stateflow for modeling, Gene-Auto for transformation, gcc for compilation, Sysgo CODEO as a hypervisor onboard OS, Eclipse and Sysgo CODEO to simplify the architecture and the configuration of the SNG router software and finally qemu-x86 to emulate the onboard target hardware.

Now that the methodological process for the development of the SNG router has been laid out, we can go into more depth about implementing its functionalities as well as presenting the different modelings and design choices that are associated with it. These different points are detailed in Chapter 3.

3

Implementing the Prototyping Methodology to Develop a Next Generation Avionic Router

This final chapter will present a case study to illustrate the advantages of using the rapid prototyping method described in the previous chapter. To this end, we decided to develop a next generation aeronautical router (Secure Next Generation (SNG) router) that transports different types of information securely and safely: communications between networks onboard the aircraft as well as communications from different ground networks (air traffic control, airline and even passenger Internet). We will describe how the methodology has been used to produce different models for routing, filtering or securing the flows. We will provide details about the production of software source code from these models and will emphasize the benefits of models in the final validation phase. Finally, to illustrate the advantages of our methodology with regard to developing this practical example, we will present the evaluation results for the performance of our avionic router and will demonstrate that the prototype performs extremely well, enabling it to process all types of aeronautical traffic until 2020 and beyond.

3.1. Introduction to next generation aeronautical communication domains

3.1.1. *Avionic and aeronautical network domains*

Aeronautics uses two terminologies to classify networks according to their use. The terminology ICAO, defined in Appendix 10 [OAC 07], is used by Airbus and is centered on communications between the aircraft and facilities on the ground. It is perfectly adapted to classify ground–board communications. ARINC 664P5 [ARI 05b] offers an alternative terminology, used by most manufacturers, that is more centered on onboard systems and the services they provide. It is thus adapted to classify onboard systems and networks.

Appendix 10 of the International Civil Aviation Organization (ICAO) [OAC 07] defines four service and aeronautical communication classes. *Air Traffic Service Communication* (ATSC) groups together *Air Traffic Control* (ATC) reserved for control, *Position Reporting* for surveillance and *Flight Information Services* (FIS) primarily devoted to navigation. *Aeronautical Operational Control* (AOC) groups together the communications required during the different phases of the flight for the safety of the flight, its dependability and its effectiveness. These two classes are used by civil aviation authorities. *Aeronautical Administrative Communication* (AAC) is used for the commercial aspects of the flight; this class is reserved exclusively for the use of airline companies. Finally, *Aeronautical Passenger Communication* (APC) is devoted to passenger communication (Internet in the cabin, telephones, etc.). Together, these four domains form an architecture called the *Aeronautical Telecommunication Network* (ATN), which includes ground–ground, ground–board and avionic (board–board) communications.

The Airlines Electronic Engineering Committee (AEEC) workgroup for the company ARINC posited an alternative terminology with four avionic communication domains, similar to that of the ICAO, in [ARI 05c]. The *Aircraft Control Domain* (ACD) groups together the onboard systems required to manage the aircraft and the flight. The *Airline Information Service Domain* (AISD) groups together all the systems that are "non-essential" for the safety of the flight, optimizing all flight processes; these systems are destined for maintenance and for the crew, including commercial personnel. The *Passenger Information Entertainment Services Domain* (PIESD) is devoted to onboard networks and systems that passengers can access from their seat. Notably, it contains *In-Flight Entertainment* (IFE), enabling passengers to watch films, listen to music, play games, read online newspapers, etc. Finally, the *Passenger-Owned Devices Domain* (PODD) groups together networks and systems, enabling passengers to use their personal communication systems (cell phones, smartphones, tablets, laptops, etc.).

Figure 3.1 shows these two terminologies in addition to the gateways between them. Although defined differently, both ATC and ACD are associated with extremely high security, safety and availability requirements. Moreover, systems and networks associated with ACD use and transmit information generally associated with ATSC. There is therefore a strong connection between ACD and ATSC and even a certain degree of confusion. The same is true for the couples AOC/AISD, AAC/AISD, APC/PIESD and PC/PODD.

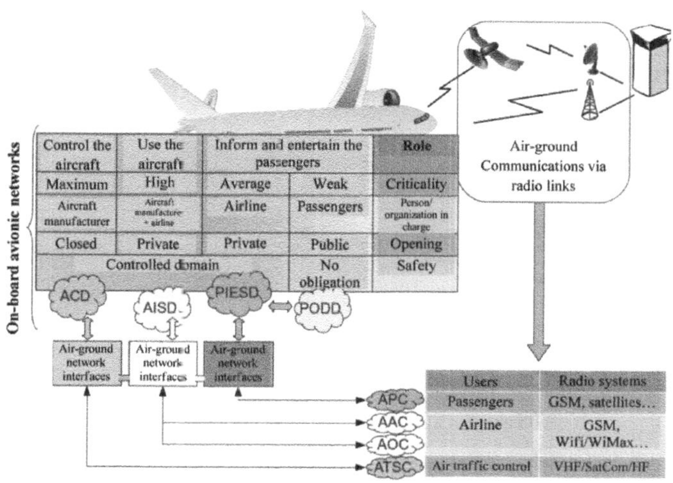

Figure 3.1. *Avionic and aeronautical networks*

Initially based on closed proprietary protocol networks, the ATN architecture validated by the ICAO is currently evolving to replace the proprietary protocols with other protocols commonly used in fields outside that of aeronautics, particularly protocols standardized by the *Internet Engineering Task Force* (IETF) that publishes and organizes the standards and propositions as *Request for Comments* (RFC). This next generation architecture is called the *Aeronautical Telecommunication Network with Internet Protocol Suite* (ATN/IPS) and its main advantage is its compatibility with public network equipments known as *Commercial Off-the-Self* (COTS). Our SNG router is entirely compatible with the ATN/IPS architecture, as illustrated in Figure 3.2. The figure shows both link layer protocols and higher level protocols, which will be discussed in the next section.

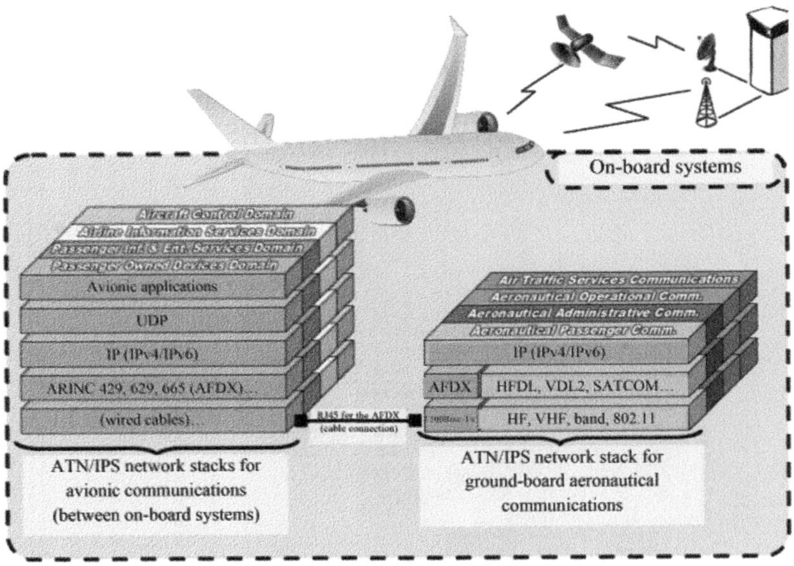

Figure 3.2. *ATN/IPS network stacks*

3.1.2. *Communication standards and protocols for next generation aeronautics*

Onboard networks primarily interconnect onboard equipment, but they also transmit data from ground networks. The links between ground networks and onboard networks are realized via a third category of network: ground–board networks. These networks use radiowaves to transmit data. This section aims to present the different technologies that are currently in use to ensure the link level between the aircraft and the ground.

Historically, the first technology used the high frequency (HF) radio band (see [DO 76]), which transmitted pilots' voices to the controllers and vice versa. Although it has a very large range (5,000 km), the HF band (1.5–30 MHz) allocated to aeronautics by the International Telecommunications Union (ITU) (see [IUT 13]) has its drawbacks: poor sound quality leading to transmission difficulties, a small number of channels that can be used simultaneously, etc. In populated areas on the ground, the *aeronautical very high frequency* band (VHF band [DO 79], from 118 to 137 MHz since 1994), which has a smaller though better quality range, quickly replaced the HF band.

At the same time, satellite technologies were being developed giving aeronautical communications the same advantages as the VHF band but for any area (oceans and land, only the poles have bad satellite coverage). Inmarsat [INM 13] was founded in 1979 and privatized in 1999. The organization offers 600 bps–10.5 kbps satellite throughput. The 66 satellite low earth orbit constellation Iridium [IRI 13] has been authorized since 2011 by the Federal Aviation Administration (FAA) for the transmission of critical aeronautical data; Iridium provides a throughput of 2.4 kbps. However, prices and communication delays by satellite are generally higher and longer, respectively, than that provided by VHF. The result is that in flight, pilots prioritize the use of

VHF, then SATCOM links when it is not possible to use VHF and, finally, HF in the event where the two other systems cannot be used.

These technologies are currently used by ATC for vocal communications. They have also more recently been used by ATC to communicate aeronautical data. The Datalink networks can therefore use the HF radio band with a link-level protocol called the *High Frequency Datalink* (HFDL) which has a maximum throughput of 1.8 kbps [DO 00]. Datalink on VHF is similar. There are four versions of the VDL protocol (VHF Datalink), but the most commonly used protocol today is the VDL mode 2. Defined by the ICAO in 1996, it offers a throughput limited to a theoretical maximum of 31.5 kbps [DO 05a, DO 12a]. Finally, SATCOM links can also transmit data. At a higher protocol level, the format of messages exchanged using these systems is called the *Aircraft Communication Addressing and Reporting System* (ACARS) and has been standardized by ARINC norms 618 to 623 [ARI 06a, ARI 09, ARI 12a, ARI 01, ARI 05a].

New technologies are also currently being researched. Indeed, establishing an *ad hoc* network where each aircraft is a node in an overall network would enable data to be gradually transmitted from the transmitting aircraft to a receiving station on the ground through the use of frequency bands that, rather than enabling the source and the destination to be directly interconnected, enable them to interconnect via a chain of aircraft relaying the data (for more information, see [BES 13]).

WiFi and WiMax are other technologies currently being researched [MAT 11, BES 13]. They have small ranges (approximately 10 km for WiMax and 100 m for WiFi), smaller than that of HF (approximately 5,000 km). They are, therefore, intended for communications when the aircraft is at the airport. These technologies could, for example, enable updates to aeronautical and weather maps to be downloaded when the aircraft is docked.

However, whatever means of communication is used, it concerns the link level (level 2 of the OSI model). As such, our SNG router, which manages the network level (level 3), does not need to know which link network is used: it sends data to the adequate equipment which consequently processes the data. The only property that can have an impact on the integration of the router within the aircraft is the throughput offered by the lower level medium: each system offers a throughput that is its own, as summarized in Table 3.1.

Name	Frequency range	Range	Data throughput
HFDL	118 – 138 MHz	~ 250 km	< 32 kbps
VDL2	1525 – 1661 MHz (band L)	∞	< 12 kbps
WiMax*	2 – 66 GHz	~ 50 km	100 kbps to 22
Ad-hoc networks*	Can use band L	≤ 600 km	from kbps to mbps

* Predictions for these technologies (they are still being developed)

Table 3.1. *Aeronautical ground–board link protocols*

3.1.3. *The benefits of a next generation avionic router: secure next generation router*

3.1.3.1. *Interconnecting avionic domains with a next generation router*

For recent aircraft, such as the A380, the civil aviation safety requirements depend on the criticality of the domain, ACD, AISD, PIESD and POD. To guarantee high levels of assurance, the current solution is based on segregation. As such, the aircraft embeds several independent avionic networks, each of which focuses on one domain or one subdomain. Historically, information such as the ground temperature and the local time, which can be accessed by pilots, could not be electronically accessed by commercial onboard staff (air stewards and hostesses) and passengers: the staff provided this information in announcements.

However, new needs have arisen: some information must now pass from the network of one domain to the network of another domain. It is, therefore, now possible to display the altitude of the aircraft (from the ACD) on passengers' monitors (connected to the PIESD). This is possible through the use of systems that interconnect the avionic networks of different domains. Opening up networks that were until this point closed poses new security and safety problems. The A380 and the B787, for example, open up networks albeit in a very restricted and very limited fashion: the ACD can send data toward the AISD via a piece of equipment called the "data diode". As its name suggests, this equipment enables, post-filtering, some data to be sent solely from the ACD to the AISD, blocking any data to the ACD. Likewise, a data diode connects the AISD to the PIESD.

Data diodes are, in fact, the first step toward opening up networks. However, the fact that they need to be designed and configured on a case-by-case basis prevents them from being scaled up effectively. For this reason, ARINC 811 [ARI 05c], which puts forward an architecture for onboard information systems for aircraft of the future, advocates an avionic architecture that is no longer centered on the coexistence of weakly coupled networks, but rather on a set of networks that are interconnected by components called routers, as illustrated in Figure 3.3.

Figure 3.3 is not a perfect reflection of research such as ARINC 811: research of this type imagines that there will not only be one router, but several pieces of equipment to improve protection against breakdowns. The underlying concept of domain interconnection remains unchanged.

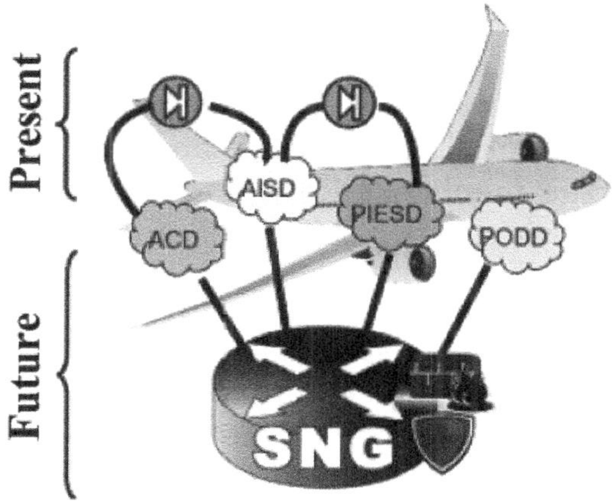

Figure 3.3. *The interconnection of avionic networks by an onboard router*

The massive deployment of AFDX, based today on the use of the IPv4 protocol as a network protocol and predicted to support its successor the IPv6 protocol, enables us to imagine the use of an IP router. The latter has existed for a long time outside the world of aeronautics: the technology has thus already been tested. Its integration into aeronautics does, however, pose countless difficult questions regarding the certification and evaluation of these systems: due to high safety and security restrictions, no certified IP router currently exists and there is consequently nothing to embed. One of the objectives of my thesis was to study the restrictions imposed on the realization of a certified router for aeronautics and propose adapted solutions to reduce costs.

3.1.3.2. *Mutualization of ground–board aeronautical links by a next generation router*

Aeronautics has a second requirement: multiplexing ground–board flows. Avionic networks are currently closed and do not communicate between themselves. Ground/board communications are duplicated for each network domain stretching from the aircraft to the ground, thereby guaranteeing the availability of the link and dedicating all the capacity to the domain using the link. This leads to an increase in the number of links and equipment (as well as associated costs).

One solution consists of mutualizing the ground/board link when the technical systems and the capacity of the link allow for it. The link then simultaneously transports several flows from different domains. This is known as multiplexing the data flow on a link. Multiplexing is ensured by the router when the data coming from different network interfaces are sent toward the onboard telecommunication equipment via one of the router's network interfaces. Demultiplexing, a reverse process of multiplexing, is then conducted by one or several ground routers.

Figure 3.4 shows an example of the use of air–ground mutualization. An onboard router is physically connected to three avionic networks: it has one link with the ground to transmit the data from all three networks. On the ground, a first router separates the passengers' uncritical data and ensures that there is a connection with the Internet Service Provider (ISP). The avionic flows dedicated to the commercial use of the device are transmitted to a second router on the ground that ensures the multiplexing of the flows and routes them toward the adequate equipment of the airline using the aircraft. ARINC 811 and its complement ARINC 821 [ARI 08], which increases problems relating to safety, advocate the use of tunnels to transport data while maintaining segregation, as shown in Figure 3.4. These tunnels must be secured due to the potential sensitivity of the information they transport: commercial data, personal data, etc.

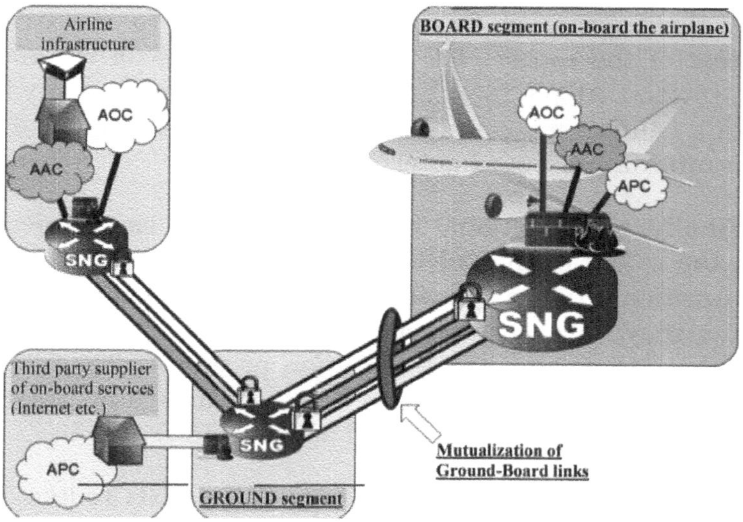

Figure 3.4. *Mutualization of aeronautical links*

3.2. Implementing the SNG router

3.2.1. *Architecture of the SNG router software*

The router must ensure several functions (explained in section 3.1.3). In fact, the router must filter and route the IPv4 and IPv6 packets between its different network interfaces: the "routing" and "filtering" functions constitute the first group of functionalities.

The router must then be able to ensure the security of communications by guaranteeing the confidentiality and integrity of exchanges via secure channels, once the end nodes of these channels have been authenticated. To limit a proliferation of rarely used secure channels and thus prevent wasting network resources, the router can be configured to mutualize these channels and thereby secure several data flows in one channel. The second group of functionalities comprises the following functions: multiplexing and securing data.

A third group of functionalities relates to the avionic context of the router. The network interfaces of the router are, in fact, AFDX interfaces: Ethernet interfaces to which certain functional requirements have been added to guarantee an improved determinism.

Finally, non-functional requirements such as the security and the safety of the "SNG router" concern the system and its functionalities as a whole. Although forming a fourth group, they have an impact on all the other functionalities and the design and implementation processes. They do not necessarily come out as code or as models and are thus processed from a methodological point of view.

Figure 3.5 shows the architecture of the SNG router. This architecture formalizes and "fixes" the decomposition of the functionalities into three classes detailed below. The functional decomposition advocated in the first stage of our methodology enables us to focus on each class in the subset of functions to be modeled, by abstracting the subsets associated with the other classes, thereby simplifying the modeling.

The Pfr class is associated with packet routing and filtering functionalities: it is thus strongly connected to the network-level protocol. Given that the router is destined to interact with both the IPv4 protocol and the IPv6 protocol, two variants have been modeled: one for the IPv4 called the Pfr4 and its counterpart Pfr6 for the IPv6 protocol.

The Pfr4 and Pfr6 classes present a shared interface with regard to the other partition classes in the system. As such, the abstract Pfr partition class defines a shared interface linking the inputs and outputs of the Pfr routing and filtering partition class with the inputs and outputs of the Piface interface partition (this will be described later). The heritage mechanism is used to derive the Pfr4 and Pfr6 classes from the abstract Pfr superclass: these two classes therefore inherit a unified input/output interface.

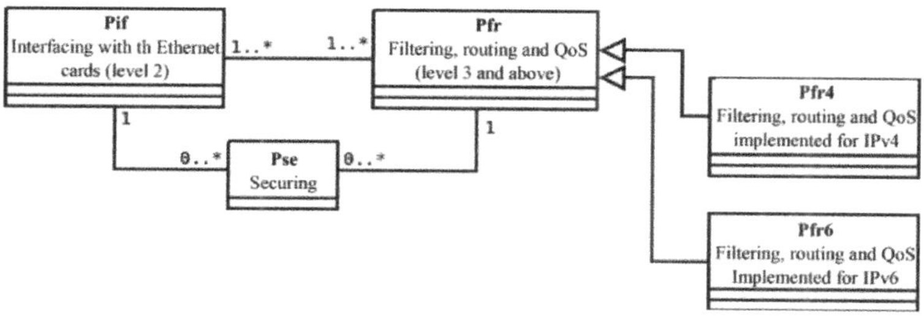

Figure 3.5. *Class diagram of the SNG router*

However, although IPv6 is defined as a replacement for IPv4, these protocols are too different for mutualizing the implementation of the Pfr4 and Pfr6 classes. We therefore did not use heritage to further mutualize the design of these two partition classes.

The Piface class links the data managing classes (Pfr4 and Pfr6) to the hardware inputs/outputs. In practice, this link consists of capturing the Ethernet frames received by network cards, extracting the data encapsulated in these frames and transmitting these data to the adequate class (Pfr4 for IPv4 packets, Pfr6 for IPv6 packets, or rejecting the invalid frames and unknown protocols). This link is bidirectional, i.e. Piface also ensures that IP packets are transmitted into the physical network.

Finally, functionalities relating to securing data by the router are implemented by a dedicated partition class, the "Partition for security" (Pse) class.

Partition classes communicate via the associations represented in the class diagram in Figure 3.5. These associations result in the implementation of message queues, managed by the IMA compatible operating system. We arbitrarily chose to use a queue length of a maximum of one message per queue with static priority pre-emptive scheduling so that instances of the Pfr class could be prioritized over instances of the Pif class. Upon arrival, packets are therefore processed by the software part of the SNG router and potentially put into the hardware queue of the Ethernet card (the network card constructor defined a static output buffer limited to 256 Ethernet frames). These choices enable the SNG requirements to be validated, as will be discussed in the next section.

The following section will describe the main functionalities implemented for the SNG router using the rapid prototyping methodology presented in this book. For reasons of balance, we decided to only illustrate the functionalities implemented for the IPv4 stack. The readers should be aware of that for the SNG router to be completely functional, the same work was required for the IPv6 protocol stack. What is more is that we will only go into detail about the routing and filtering functionalities.

3.2.2. *Routing functionalities (IPv4)*

3.2.2.1. *The principles of routing for the SNG*

The first and primary function of a router is to route the data between the different networks the router interconnects. As discussed in section 3.1.1, the SNG router can be connected, in particular, to the ACD, AISD and PIESD network domains. All these domains use the AFDX as link layer (level 2). The network interfaces of the SNG router are therefore AFDX interfaces.

These network domains currently use the IP version 4 protocol (initial standard published in 1982 by the IETF in the RFC 791 [POS 81a] for the network layer (level 3)). However, aeronautics is already preparing to use its successor, the IP version 6 protocol (IPv6, initial standard published in 1995 in the RFC 1883 and updated in the RFC 2460 [DEE 95, DEE 98]). The IPv4 and IPv6 protocols can coexist in the same hardware infrastructures. They both decompose data into finite subsets containing series of bytes, known as "packets".

After lengthy discussions with Thales Avionics, we decided to focus on a *dual-stack* router: this type of router contains several stacks of networks, each of which is dedicated to a unique level 3 protocol. For the SNG router, this means independently implementing mechanisms for IPv4 and others for IPv6. The identification of which mechanisms to use is conducted using the information transmitted by the link layer: for each packet received, the link layer indicates whether mechanisms for IPv4 or IPv6 should be used.

Packet routing is realized with a routing table: this indicates to which equipment and therefore via which network interface each received packet should be sent. For safety reasons, we decided not to use the dynamic method to create and manage this routing table. In the SNG router, the routing table is a static configuration element that is defined when the router is installed in the airplane. It cannot be modified later.

Higher level layers (layer 4: transport layer, layer 5: session layer, layer 6: presentation layer and layer 7: application layer) have no impact on the router: data and metadata are included in the IP packets and the SNG router routes the packets atomically by only modifying what is required by network standards. Therefore, the modification of *time to live* (TTL) and the incremental calculation of the checksum were specified in the standards so that there is no impact on the data transported in the packet.

Avionic networks currently use IPv4 packets that are transported by the AFDX and that contain UDP datagrams (transport protocol defined in 1980 in the RFC 768, see [POS 80]). Future requirements that are moving toward the IPv6 can be part of either the UDPv6 (UDP transport protocol modified by the RFC 2460 [DEE 98]) or the TCPv6 (*Transmission Control Protocol for IPv6*, TCP transport protocol published in the RFC 793 [POS 81b] in 1982 and modified in 1999 by the RFC 2460 so that it can operate with the IPv6 protocol). These considerations help guide the evaluation of performance, presented at the end of this chapter.

The initial idea of making the SNG router a gateway between IPv4 and IPv6, converting the packets from one network protocol to the other while they are routed, was abandoned: gateways are in their very nature vulnerable (as the number of RFCs relating to solutions demonstrates: [AOU 07, HUI 06] as well as [DES 10]) and the risks they pose to security and safety have been considered too high in comparison to the expected benefits.

3.2.2.2. *Implementing routing functionalities*

Given that the current avionic domain is based on static and well-established networks, we implemented a routing mechanism that uses a static routing table. The latter is defined during the configuration phase and is non-modifiable when the router code is running. This table (its structure is presented in Table 3.2) is used by the state machine for routing IPv4 packets (illustrated in Figure 3.6). This state machine diagram was modeled using Stateflow.

3.2.2.2.1. Description of the structure of the routing table

The routing table is a table of N routing entries, where N is a constant numeric value chosen during the design phase according to the needs. A low value limits the maximum number of routes processed by the router, but saves the memory. Given that the routing table is browsed linearly to route the packet, a high value can damage performance and consume the memory.

We chose N = 32 for our experiments: this value is high enough to contain all our routes. The amount of memory taken up by routing is 12*N, or 384 bytes in our case.

Each routing input, described in Table 3.2, is a basic structure comprising the address assigned to the router in the network, the network mask, the partition output port toward which the packets corresponding to the input and the associated *maximum transmission unit* (MTU = maximum size of the fragments that are authorized as output for this route) are directed.

Field name	Offset	Size	Description
Address	0	4	Router address in the destination network
Mask	4	4	Sub-network mask associated to the address
Port_Dest	8	1	Interface number, through which the packet is transmitted
MTU	9	2	Accepted packet size
(reserved)	11	1	(padding byte)

Table 3.2. *Structure of the static routing table*

If a packet larger than the MTU is received by the router, the packet is considered to be invalid, thus deriving from a faulty system, and the packet is therefore silently rejected.

A padding byte was added to align the routing table entries in multiples of 12, thus optimizing memory access to our architectures (currently 32 or 64 bits).

3.2.2.2.2. Explanation of the implementation of routing

Figure 3.6 represents the state machine that routes the IPv4 packets received by the SNG router. The router iteratively tests each route, the first correspondence leading to associated behavior, *Forwarding* or *Local_Processing*, both of which are defined below.

If there is no corresponding route, the packet is dropped. The signal *No_Route_Found,* shown in the diagram, is therefore triggered. This serves to raise an internal variable that can be used for debugging purposes.

If the packet's destination address corresponds to the router address in the routing entry, the state machine for routing redirects the packet toward the *Local_Processing* subsystem. This subsystem enables the SNG router to respond to a potential ICMP ECHO request [POS 81c, CON 06], the only local service our router implements to facilitate the verification and validation of the SNG router for its configuration and installation in the aircraft.

The network address associated with the route is obtained by masking the router address with a mask. The address of the packet's destination network is obtained by masking the destination address of the IP packet with the mask of the routing entry. When these two network addresses correspond, the packet is routed toward the output port via the *Forwarding* subsystem. The MTU is taught to reject the packet if necessary. A potential fragmentation system could use this data to determine whether it is necessary to fragment the packet or not; however, the avionic context meant that this possibility was unnecessary.

Multicast packets (packets whose IP destination address is greater than or equal to 224.0.0.0) or local loop packets (whose destination address is 127.x.y.z) are systematically rejected, as are general diffusion packets (whose destination address is composed of the network address followed by bits all at 1, for example, 10.255.255.255 for the network 10.0.0.0/8) and packets destined for the network (the destination address of the packet is a network address).

Packets must not be directed toward an output port that is the same as the input port so as to avoid impersonation attacks called *IP-spoofing*. They must not be directed toward a port associated with a deactivated network interface (for instance, when the cable is physically unplugged, which is indicated by the input *enabled_in_ports*).

As detailed in the two previous sections, the conditions for a refusal to route result in the packet being rejected. Indeed, a valid packet in terms of avionic restrictions cannot correspond to one of these conditions, and the router therefore rejects the packets matching any of these conditions as invalid.

The graphic approach of the models used with our methodology is short and complete (the entire routing algorithm fits in one page): it enables the model to be reread quickly. The model that we created in Figure 3.6 visually summarizes the set of cases for routing packets and the processing operations realized.

3.2.3. *Filtering functionalities (IPv4)*

3.2.3.1. *The principles of filtering for the SNG router*

The router connects networks of different criticalities. There is a difference between attack vectors in the ATSC, which is only accessible to surveillance and air navigation services via certified equipment, and the APC, which is accessible to passengers via their personal communication devices. The impact of an attack also varies: the unavailability of the APC is an annoyance for passengers and may have an impact on the airline's brand but has no effect on the safety of the flight, whereas the unavailability of the ATSC can lead to losing control of the aircraft, the plane crashing and the loss of life.

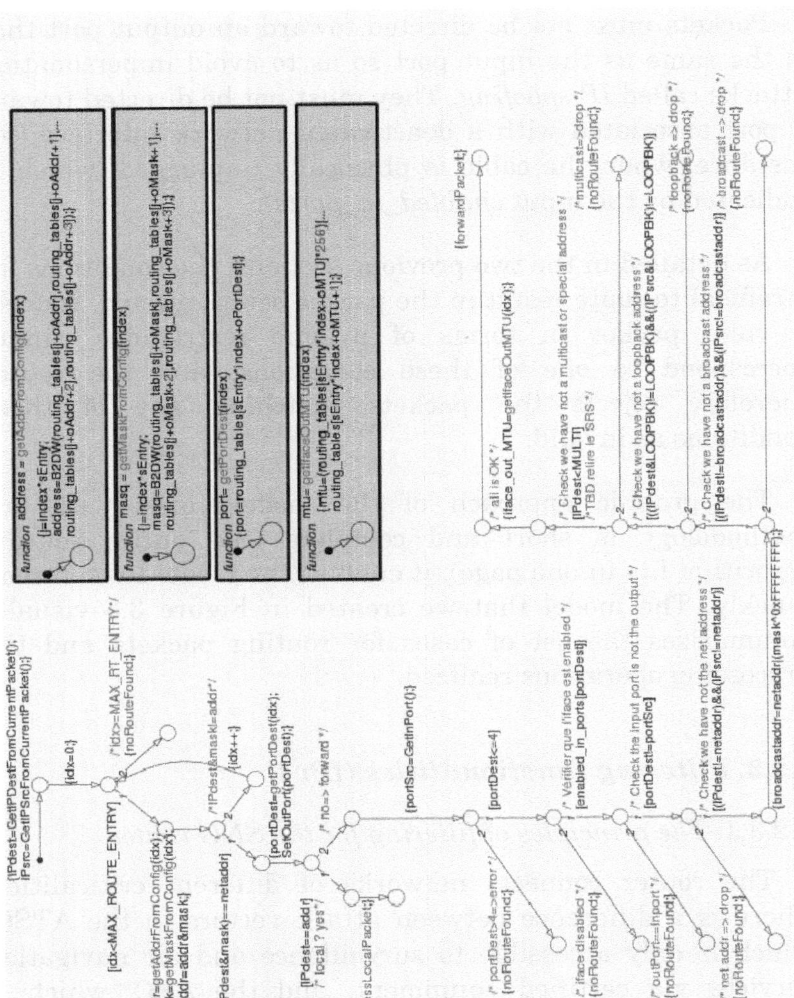

Figure 3.6. *State machine diagram for routing IPv4 packets*

Isolated from one another until now, the interconnection of networks increases their vulnerability and thus requires systems to be implemented to resolve it. However, high safety operation restrictions make it difficult and even impossible to implement complex software or hardware solutions in systems such as these embedded routers.

After reflecting on this point, we implemented a static mechanism that filters the IP header of each packet received before any other processing operation is realized. We imagined including the most advanced mechanisms such as *deep packet inspection* (for more information, see [DUB 03]), which analyzes both the IP packet and its content, i.e. the protocol headers of the higher levels and the application content. These mechanisms are independent of the filtering of the IP headers mentioned above; we therefore decided to develop two complementary systems that are deployed in the avionic architecture rather than one complex system that is complicated to implement, certify, evaluate and maintain.

3.2.3.2. *Implementing filtering functionalities*

The Pfr4 class not only enables IPv4 packets that pass through the router to be routed, but also offers a basic stateless filtering functionality. This filtering verifies the IP header of each IPv4 packet entering the router independently of the history of the previous packets routed (or dropped) by the router.

Filtering is realized by the state machine diagram Filters, modeled using Stateflow and represented in Figure 3.7. This state machine takes 192 configuration bytes and gives a decision to either accept (the packet can continue being processed) or drop (processing stops, the packet is silently rejected) the packet.

Following the example of the routing table, defined at configuration and essential for IP routing, IP filtering is based on a filtering table comprising M entries with 48 bytes each. The structure used is represented in Table 3.3.

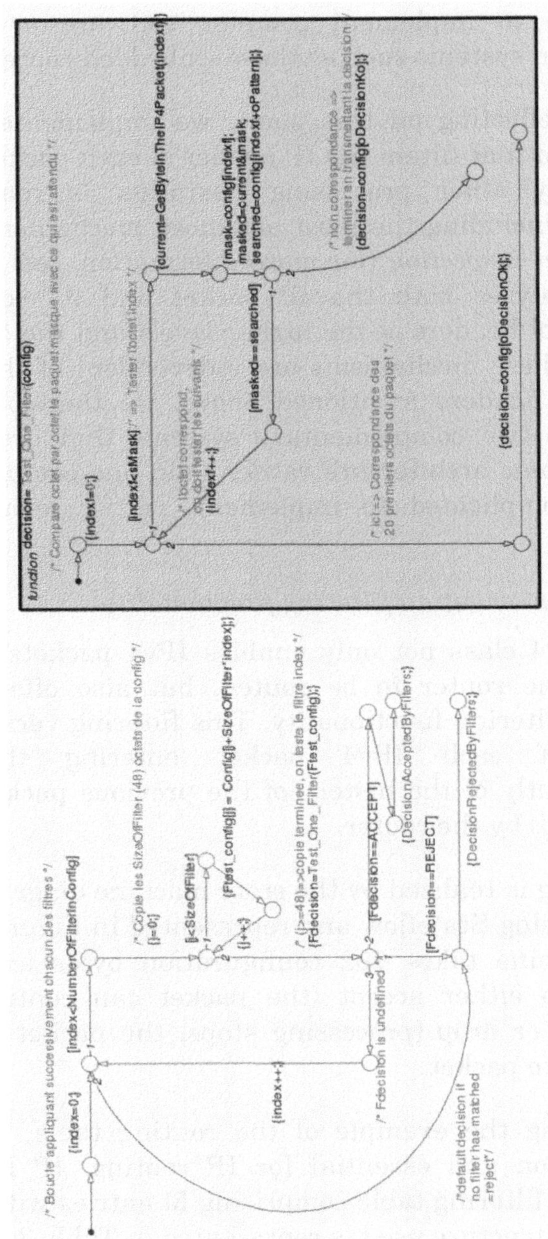

Figure 3.7. *State machine diagram filters*

[index]	Content of filtering entry
[0] accept the packet if Ipdest==10.1.2.3./32	Mask= { 0..0, (ipdesk), 255, 255,255,255,0..0 } Pattern = { 0..0, 10, 1, 2, 3, 0..0 } Decision_if_found = Accept Decision_if_not_found = Continue
[1] drop the packet if Ipdest=10.0.0.0/8	Mask= { 0..0, (ipdesk), 255, 0,0,0,0..0 } Pattern = { 0..0, 10, 0, 0, 0, 0..0 } Decision_if_found = Drop Decision_if_not_found = Continue
[2] drop the packet if TTL > 6	Mask= { 0..0, (ttl), 240, 0..0 } Pattern = { 0..0, 0, 0..0 } Decision_if_found = Continue Decision_if_not_found = Drop
[3] accept any other packet	Mask= | { 0..0 } Pattern = { 0..0 } Decision_if_found = Accept Decision_if_not_found = -

Table 3.3. *Example of a filtering table with four entries with two conditions*

Given that four filters meet our requirements, we chose M = 4. However, a version with more than four filters can easily be derived by modifying the NUMBER_OF_FILTERS constant in the model for the Pfr4 class and then recompiling the development project for the SNG router.

Filtering consists of iteratively applying each available filter to the IP header. The decision applied is one of the following:

– accept the packet, in which case the packet is immediately transmitted to the next subsystem in the processing chain;

– refuse the packet, in which case processing immediately passes to the next packet, silently rejecting/dropping the packet;

– continue processing, i.e. applying the next filter's entries until a decision is made about whether to accept or refuse.

The decision applied depends on recognition: when the pattern is recognized, the decision is indicated in Decision_if_found; otherwise, the decision is indicated in Decision_if_not_found.

Applying a filter to the IPv4 header of the packet consists of masking the bits of the IP packet that do not need to be tested and then comparing the remaining bits to the pattern that is sought. The fields Mask and Pattern in the entry are used to this effect and enable the 20 bytes of the IP header to be processed, whichever field is being filtered.

The systematic comparison of 20 bytes is an acceptable level of filtering as avionic systems do not use IPv4 options and the IP headers are thus always exactly 20 bytes.

For the purpose of illustration, Table 3.3 shows a four-filter example that unconditionally accepts the 10.1.2.3 packets, drops the packets of other nodes in the 10.0.0.0/8 subnetwork and drops the packets with a TTL higher than 16.

The default behavior of the router is to reject the packets. This behavior is applied when no filter has enabled a decision to be made. This decision was made to maximize security.

Six padding bytes are present at the end of each entry in the filtering table so that the filters can be aligned in multiples of 16: the aim of this is to optimize filtering execution time.

The state machine diagram in Figure 3.7 has a cyclomatic complexity that is higher than 1 (there are several possible paths to get from the initial state to the final state) and implementation could lead to the creation of an infinite loop between certain states.[1] Our methodology is, however, based on the use of a formally verifiable model; indeed, we verified with the modeling tool that no blockage could occur when the model is running: whatever the initial state of the system, it always arrives at a final state within a finite amount of time. Likewise, formal proving tools enabled us to verify the coherence of the types of data, avoiding and correcting cases of integer overflow.

3.3. Evaluating the performance of the SNG router

This section aims to validate the very good level of performance obtained by the software system that was developed using the rapid prototyping methodology presented in the previous chapter.

To this end, we will start by presenting the different network topologies that are envisaged before describing the different metrics enabling us to quantify the network performance of our SNG router and the measuring tools we used. The numeric results will then be presented alongside a reminder of the methodology and an interpretation of the results.

3.3.1. *Experimental framework*

3.3.1.1. *The network topologies considered*

To define the topology of an optimal test, we focused on several topologies to be installed and configured in our experiments. The legend of the topologies presented here is indicated in Figure 3.8.

1 Also known as an endless loop in which the execution passes from state 1 to state 2 and then from state 2 to state 1, thus the cycling continuing endlessly.

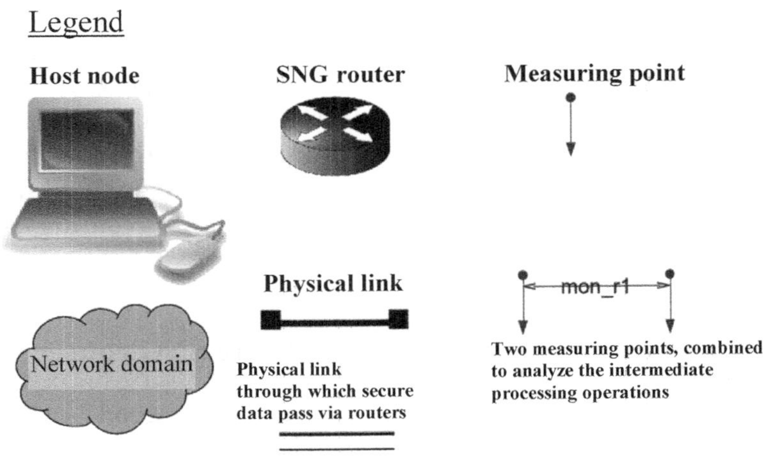

Figure 3.8. *Legend used for network topologies*

Presented in Figure 3.9, the first topology comprises a host called H1, which is physically connected by an Ethernet link to the host R1. This topology was first used to verify and validate that the implementation of certain functionalities is operating correctly, thereby enabling us to validate the response to the ping (an ICMP ECHO service) and filtering as well as to partially validate the routing functionalities of the SNG router.

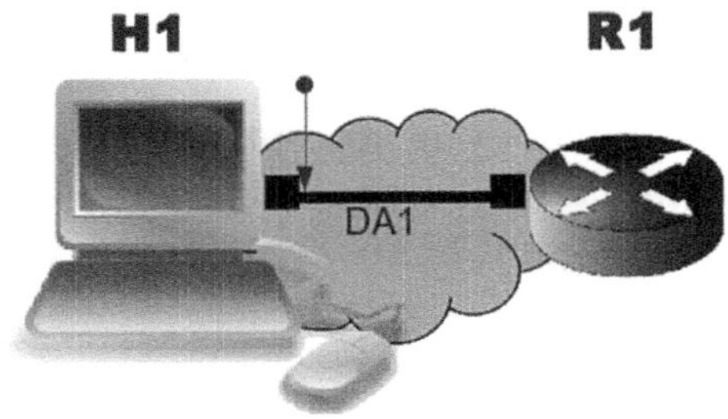

Figure 3.9. *Topology "A" with one host and one router*

However, the topology in Figure 3.10 is more adapted to validate routing. Constituted of three hosts and three logical subnetworks, it does however only contain one SNG router and does not therefore enable the security performance between SNG routers to be evaluated.

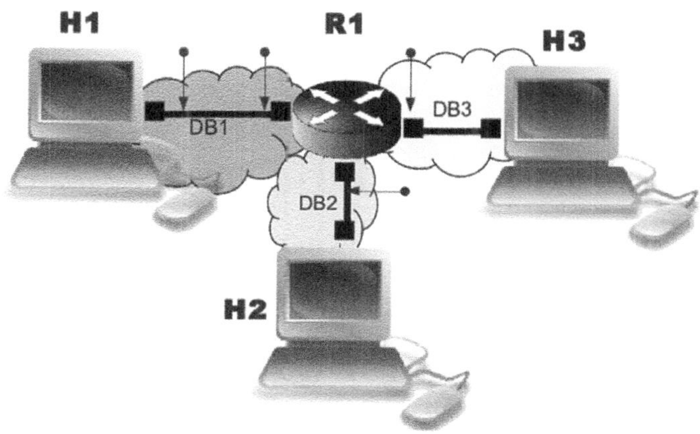

Figure 3.10. *Topology "B" with three hosts and one router*

For this reason, we favored the topology presented in Figure 3.11 which is constituted of two end hosts, H1 and H2, which serve as a source and a well, respectively, or even as a source and a mirror (sending back responses that are adapted to the source's requests) and that comprises one path crossing two SNG routers. These neighboring routers can establish, according to the configuration, a secure channel between themselves in order to validate the safety functionalities. The topology is thus symmetrical: H1 – R1 == <<symmetry>> == R2 – H2.

This topology can also validate routing: the routers route the packets between two networks that are connected to two of their network interfaces. Moreover, the configuration mutualizes the link between the two routers, R1 and R2, so that they can be simultaneously connected to four logical networks, each of which is associated with a different level of security, as presented in Figure 3.12.

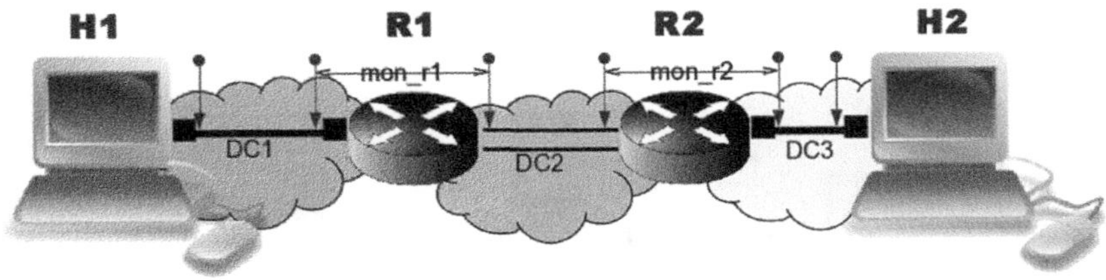

Figure 3.11. *Topology "C" with two hosts and two routers*

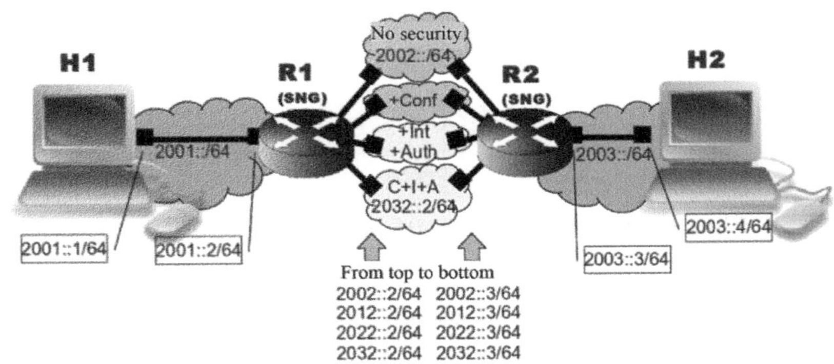

Figure 3.12. *Logical networks enabled by topology "C"*

We therefore chose topology "C", presented in Figure 3.11, to test and stress the SNG routers (see section 3.3.3).

3.3.2. *Metrics and tools for evaluating performance*

3.3.2.1. *Selected metrics*

The domain of network metrology provides us with different metrics that are measurable at different places in our network topology.

3.3.2.1.1. Processing router capacity

The first metric is the processing capacity of the SNG router, which is evaluated as bytes per second. This metric is more commonly called the "[maximum] throughput of the router" or the "operator throughput". It is measured at the output port of the SNG router. Other units can be used to qualify this throughput: bits per second (8 bits = 1 byte) or packets per second, when the packets are of a constant size.

3.3.2.1.2. Throughput obtained by users

Given that we are using topology "C" (presented in Figure 3.11), it is interesting to study the throughput between the end hosts, H1 and H2, also called the "user throughput". In this topology, a relationship connects the "user" and the "operator" throughputs: as there is one path, the user throughput is restricted by network elements, more precisely by the most restrictive element, the "bottleneck". For our example, we used 100 Mbps Full Duplex Ethernet cards and, as shown in the next section, the results show that the restrictive element is our SNG router. As such, the relationship is as follows: "user throughput between H1 and H2" = "operator throughput /processing capacity of the SNG router".

3.3.2.1.3. Router processing time

A third metric is the router's packet processing time. The time induces a latency, measured between H1 and H2, called *one way delay* (OWD). It is, however, complex to measure the OWD: it requires the times of the two end systems to be synchronized and the clock shift to be corrected online during testing.

We therefore favored the time metric *round trip time* (RTT). In fact, as our topology is symmetrical and in full-duplex mode, the RTT is thus double the OWD. This resolves the previous problem of having to synchronize the clocks.

The time difference was also overcome because of the different order of magnitudes of the RTT measurements (~milliseconds) and the accuracy of the clocks used (~microseconds, the values were measured on the observer systems).

The jitter is another metric that derives from values resulting from RTT measurements.

3.3.2.1.4. Loss rate

We also measured the loss rate of the router. This only relates to the UDP transport protocol; it does not relate to the TCP. Indeed, the latter has mechanisms to detect and correct data loss, and there is therefore no loss from the point of view of the application that uses a TCP connection to communicate.

3.3.2.1.5. Availability of the system

A final metric that we selected is the availability of the router. Most avionic systems are, in fact, supposed to function without interruption or to automatically restart if there is a breakdown that can be fixed.

3.3.2.2. *Chosen measuring tools*

Different classes of tools are available to quantify these metrics: active measuring tools and passive measuring tools.

3.3.2.2.1. Passive metrology tools

Passive tools use data forwarded through networks to extract information. Network "sniffers" are some such tools: they capture the data transiting inside the network, analyze them either quickly or thoroughly, and copy the data onto a disk for further analysis offline.

The network sniffer Wireshark [COM 13a] conducts online listening and analysis, rapidly producing results, but it consumes a great deal of memory and processing resources, which can distort the results: packets are lost because the processor is overloaded, the software is suddenly shut down by the operating system and overall instability when the RAM is saturated.

Alternatives are tcpdump [VAN 13], which we have also used, and dumpcap [COM 13b], which is used internally by Wireshark. These network sniffers listen to the network and safeguard the data captured in files.

Finally, we used the program socat [RIE 11], nicknamed the "new Swiss army knife of the network", which offers countless opportunities for packets relay: it relays application data, TCP data flow, IPv4 and IPv6 packets, Ethernet and gateway frames between the different protocol layers and even redirection toward files and pseudo-files. We used it for the low-cost duplication of packets toward a network interface linked to an end host dedicated to network observation.

3.3.2.2.2. Active metrology tools

The second class of measuring tools is active tools. Indeed, in some cases, no data are passed through the networks, or rather the passing data do not allow the useful and/or quality information to be extracted. In this case, active tools transmit data about the network, which can then have an impact on the network. Transmission must therefore be realized in the least intrusive way possible for the network.

As such, the tool pathload [DOV 02, DOV 06] transmits torrents of packets to find out the capacity of the path being tested. As these torrents pass through the network, intermediate elements can become congested and consequently lose packets. However, our measurements with this tool proved to be too variable, too unstable and unreproducible and so we looked for other measuring tools.

Another active tool that we used is iperf [GAT 13]. This tool enables the throughput of a path to be determined, in our case the throughput from H1 to H2 via R1 and R2. It can work in TCP, loading the path to the maximum, thereby indicating the maximum throughput with a TCP connection. It can also function in UDP and it transmits a constant throughput of datagrams, indicating the loss rate and the reception throughput at arrival. This tool is therefore very intrusive.

RTT measurements were conducted using the standard tool "ping6". This tool sends small "ICMPv6 ECHO request" packets with the time it was sent. When the "ICMPv6 ECHO response" packet is received in response to the request, ping6 calculates the time between the request and the response: this is *RTT*. The variance of the RTT, known as jitter, can be determined by measuring numerous RTTs. Although active, this tool does not overload the network (in our case, 104 bytes sent per second on a path with a capacity higher than 10 Mbps) and we judged its intrusiveness to have an insignificant impact on our experimental network.

Table 3.4 summarizes the different metrics evaluated and their units, in addition to the tools used to measure them.

Metric	Unit	Measuring tools
"User" throughput	{_, k, M, G} {bit, byte, packet}/s	iperf, Wireshark
"Operator" throughput	{_, k, M, G} {bit, byte, packet}/s	
RTT delay	{_,m, µ, n} second	Ping6, tcpdump
Router processing time	{_,m, µ, n} second	
Jitter	{_,m, µ, n} second	
Loss	(%)	Iperf in UDP
Availability		

Table 3.4. *Summary of the metrics and the tools*

3.3.2.3. *Use of active tools to generate traffic load*

3.3.2.3.1. The different loads considered for the network

In the previous section, we mentioned the problem of changing network performance according to whether there is data passing through the network or not. As such, measuring the router's maximum processing capacity can be underevaluated if the data flow consumes part of the capacity at the same time the capacity is being evaluated, for instance, when several intrusive tools are evaluating capacity at the same time.

However, as will be seen from the results, the group of measurements that is most impacted by the state of the network is the time measurements. While a network that is loaded with countless data flows is being measured, the packets our active tool ping6 uses as measurements are not prioritized and must therefore pass through the message queues at the entry of each system or subsystem. As a consequence, the time they take to go there and come back is strongly penalized by the queues in comparison to the same measurements when the network has nothing to transmit.

We decided to study the performance of the SNG router with three different traffic loads. In the first case, no user data circulate the network: we will call this *without load*. The empty network prevents the use of passive metrology tools, but it enables the most reliable results concerning the maximum capacity (throughputs) of the network and its components to be obtained.

From the maximum capacity measured without load, where the restrictive element is its saturation limit, the second load can be implemented: the *maximum load* or "full load". The network is, in fact, still between "without load" and "maximum load". We can thus expect that, in real usage, the performance of the SNG router be confined between the "without load" performance and the "maximum load" performance.

We also realized measurements with a third traffic load: the *provisional load* or *nominal load*. This load type represents the traffic that we researched to be the traffic in the context of avionic use for the SNG router (described at the beginning of this chapter).

3.3.2.3.2. The imagined nominal traffic load

We conducted a study to determine the types of flows that are likely to pass through the SNG router. As presented in section 3.1, three main classes of traffic are likely to pass through the networks linked to the SNG router: ATSC, AOC + AAC and APC flows.

These flows are, respectively, for highly critical systems and necessary for the flight (*Air Traffic Services*), for onboard systems managed by the airline for commercial reasons, security, regulation and flight optimization (*AOC + AAC*) and, finally, for the "passenger" flows of the future (*APC*) comprising data exchanged by passengers toward ground networks via the onboard infrastructures in the aircraft (for instance, the onboard WiFi provided by the airline).

For these classes of flows, we modeled, according to the types of protocols used (TCP or UDP), the time between packets (called Doff, in seconds) and the size of the flows (called Don, in bytes), in addition to their mathematical distributions and their parameters. Therefore, Doff is the distribution of the time intervals between the beginning of the flows and Don is the distribution of the length of the flows.

ATSC, AOC and AAC flows already exist and exclusively use the UDP transport protocol [BEN 12]. From the COCR

[COC 07], we were able to determine the minimum and maximum sizes of the IPv4 packets exchanged. Given that it has been proven that the choice of the distribution has an insignificant impact on the results [CNE 09], we decided to use the same distributions for both Dons and Doffs. We thus modeled the flows as described in Table 3.5.

UDP flows		Parameters	Air to Ground	Ground to Air
ATSC	Don: Uniform [bytes]	Min	88	88
		Max	2500	4000
	Doff: Uniform [second]	Min	0	0
		Max	2	2
AOC + AAC	Don: Uniform [bytes]	Min	90	90
		Max	5000	125
	Doff: Uniform [second]	Min	0	0
		Max	2	2

Table 3.5. *Modeling by the mathematical distributions of the ATSC, AOC and AAC flows*

Passenger flows are much more complex; so we relied on different results of Internet traffic modeling, based on the use of processes called "ON-OFF sources". Characterizing the modeling APC traffic is beyond the scope of this book. Furthermore, given that the objective of this section is to illustrate the performance results obtained by the use of our rapid prototyping methodology, we will not go into detail about modeling passenger traffic using ON-OFF processes. Interested readers are advised to consult [LEL 94] and [WIL 97] for details on this topic.

3.3.2.4. *The hardware platform used to evaluate the performance of the SNG router*

We used different systems to realize topology "C".

Onboard SNG routers ran on classic PC architectures equipped with Intel® Xeon® E5603 1.60 GHz processors. These 64 bit quad-core processors equipped with L3 caches of

4 Mo 1.60 GHz internal memory were, in fact, underused as the IMA platform system was configured to use just one core in 32 bits. Likewise, the 2 GB of the RAM 800 MHz DDR2 was underused: the software for the SNG router requires less than 2 MB at execution to operate.

These routers were each equipped with a quad-port Intel® PRO/1000 PT Gigabit Ethernet network card, a card based on two 82571 GB controllers and connected to a PCI express port (the "PCIe" port offers a minimum throughput of 2 Gbps for communications with the rest of the hardware). These cards were configured in 100base-Tx-FD mode, i.e. their transmissions took place at a speed of 100 Mbps in full duplex (or 200 Mbps by adding up the two directions). The systems were linked together by crossed category-6 RJ45 cables to limit data loss. Full duplex was chosen to avoid loss and retransmission delays caused by collisions between the Ethernet frames.

The end hosts and monitors were classic workstations equipped with Debian Lenny (5.0) systems with a 32 bit 2.6 GHz dual-core processor with 4 GB of RAM. Their network card was also manually configured there in 100base-Tx-FD; during the tests, we verified that these hosts did not cause the performance limits measured while testing.

3.3.3. *Results of the performance measurements*

The experimental framework and metrics measured have been described in the previous sections. We conducted several measurement campaigns to evaluate the different characteristics of the router according to different parameters. The exhaustive list of results is too large to be presented here and, what is more, some of the results are redundant. We have, therefore, decided to present only the most significant results to demonstrate which effects lead to the variation of this or that parameter.

In fact, performance varies according to several parameters:

– The first parameter is the transport protocol used to transmit data. We used two common protocols, UDP and TCP: the former is message-oriented and the latter is oriented at continual data stream and is the only protocol that guarantees there is no loss.

– The second parameter is the load imposed on the network during measurement. We have three loads: empty network, nominal load network (prospective avionic load) and maximum load network.

– The third parameter concerns the number of parallel flows used for the active test data. We, in fact, expected that the performance of the router would change according to whether the data was transmitted via one TCP flow or via a multitude of TCP flows. The same question applies to UDP, although the notion of flows is more difficult to determine due to the absence of an initialization and finalization "UDP flow" datagram.

– The fourth point concerns security and knowing the cost of using the confidentiality and integrity + authentication mechanisms. The last two are closely connected to the implementation of the SNG router (they both use the same signature) and hence are considered inseparable.

– The fifth point concerns the size of the IP packets exchanged.

We also questioned the costs of the position of the routing entry in the routing table.

Similarly, we verified the availability of the router over the "long period" of one week, while the other measurements took just minutes and sometimes hours.

We also studied the impact of using an emulator in place of a real system.

Let us remind you that we measured times, throughputs and other metrics.

It is worth pointing out that each result shown here is the outcome of at least five, sometimes more, repetitions of the measurement: this enabled the stability of the result and its reproducibility to be verified.

Dimension	Possible values
Metric	{RTT [s], processing throughputs/capacities [Mbps, pps],...}
Transport protocol	{TCP, UDP}
Load during the test	{empty, nominal (800 kbps), maximum (48 Mbps)}
Number of active flows	$\{1, 4, 8, 16\} \subset \mathbb{N}^*$
Security	{without secu, conf., integ + auth., conf.+integ.+auth.}
IP packets length	$\{60, 1200, 1500\} \subset \mathbb{N}^*$
Position in the routing table	$\{1, 2, 3, 4, 5\} \subset \mathbb{N}^*$
Duration of the test	from 60 seconds to 7 days
Repetition of the test	$\geq$ 5 times
Hardware	{real (embedded target), emulated (qemu)}

Table 3.6. *The dimensions of the space covered during evaluation*

We summarized these independent parameters in a 10-dimensional hypercube (described in Table 3.6). The hypercube was browsed from the origin to different vertices, thereby constituting the total set of evaluation results for the performance of the SNG router. Detailing all these results is not necessary for the purpose of this book. We will therefore concentrate on the most representative values to illustrate the very high level of performance obtained by the SNG router software which was developed through our model-based methodology for rapid prototyping complex onboard systems.

3.3.3.1. *Impact of the number of flows processed by the SNG router on its performance*

We will first present the capacity measurements conducted in UDP. In addition to the measurements realized with one UDP flow that saturates the processing capacity of the SNG router (the equivalent of saturating the path between H1 and H2, given that the SNG router represents the bottleneck of the network topology we are discussing in this section), we conducted the experiment again using 4, 8 and then 16 parallel flows.

Each time we reported the total quantity of data received by the H2 wells, i.e. the total reception throughput for the 4, 8 and 16 simultaneous flows (see Figure 3.13).

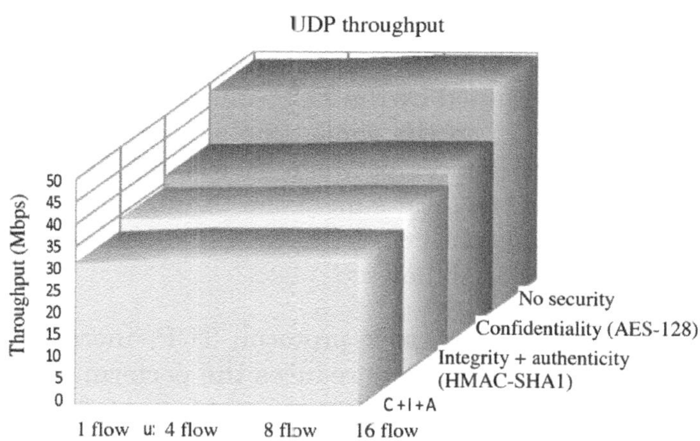

Figure 3.13. *SNG router capacity with several secure UDP flows*

In UDP, flows coexist but are independent. As such, increasing the number of flows does not lead to an increase in the reception throughput. Figure 3.13 therefore looks like a staircase with thin steps.

Enlarging the differences shows that the increase in the number of flows results in a slight, but insignificant, increase of throughput (up to +2% for 16 flows). We reproduced the experiment without the SNG router, enabling us to confirm this trend which is probably due to a better management of the message queues exiting the H1 network interface.

Another important point concerns the interactions of the effect of multiple flows with security. In fact, Figure 3.13 shows that these two parameters are independent. This is explained by the implementation of the security functionalities of the router, which handle IP packets independently of their content; the SNG router does not work at layer levels higher than that of the network layer (3) and does not therefore differentiate between flows from the same logical host (same IP address for the flows).

We then conducted the experiment again and measured the throughput attained by the TCP connections. Figure 3.14 presents the results on the same type as graph as before to emphasize the differences between TCP and UDP. We once again represented the total sum of the throughputs, which is more useful than the throughput obtained by a TCP connection for 4, 8 or 16 flows.

For this second transport protocol, TCP, increasing the number of flows significantly reduces the performance of the SNG router. As such, between the extremes of 1 and 16 flows, the total TCP throughput is reduced by 15%. This reduction appears to be exponential on the curve as the scale for the number of flows is 1, 4, 8 and 16. We do not, however, have enough information to qualify this (exponential, linear, logarithmic and parabolic) reduction.

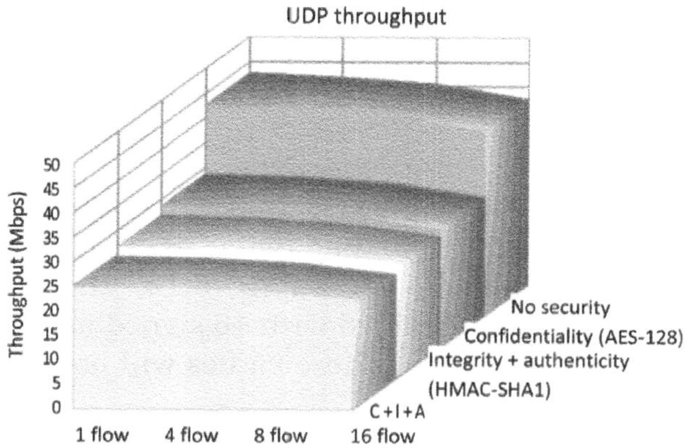

Figure 3.14. *SNG router capacity with several secure TCP flows*

The reduction of throughput upon the increase of the number of flows is explained by mechanisms for regulating TCP throughputs. These mechanisms are the most effective when the TCP connection is the only one using the path. When several connections coexist in the path, they mutually deteriorate as the congestion caused by a TCP connection also has an impact on all the other connections, thus reducing all their throughputs.

Sharing measurements for network path capacity is the objective of research in the community; it has been established that certain versions of TCP are not *fair* in comparison to others. We used Cubic TCP and it generally has a good *fairness* score [SAN 08].

In both TCP and UDP, it can be stated that securing is independent of the number of flows, and there is homothety between the curves; whatever securing mechanism is used, increasing the number of flows tends to reduce the capacity of the SNG router. Moreover, although there are many flows, securing tends to reduce the capacity of the SNG router. For instance, the encrypted values correspond to the no security reference values which are multiplied by a ratio of 0.70

(corresponding to a reduction of 30%); likewise, for other securing modes, values with signature are multiplied by 0.65 and values with encryption and signature are multiplied by 0.57.

It is important to emphasize that all these throughput measurements are higher than what is required by the aeronautical industry (see Table 3.1): the minimum throughput measured obtained with 16 signed and encrypted TCP flows equals 22 Mbps. These values will be discussed in detail in section 3.3.3.4.

3.3.3.2. *Impact of the securing mechanisms on performance*

To summarize the impact according to the metrics used, we decided to group together the main results from Table 3.7. This shows that if trends are the same regardless of the metric, the reduction of the values remains specific to the metric used. As such, the implementation by our SNG router of the encryption algorithm AES-CBC-128 is less costly in terms of calculation time than the signature algorithm HMAC-SHA1. Tunneling is, however, the most used calculation regardless of the safety services used.

[100% reference = no securing]	Throughput (byte/second)		RTT	Throughput (packets/second)	
	UDP	TCP	ICMP	UDP (60o)	UDP (1200o)
Confidentiality: AES-CBC-128 encryption	-7.5%		-2.5%	-1.6%	-5%
Integrity and Authentification: HMAC-SHA1 signature	-14%		-4.5%	-2.8%	-10%
Tunneling: encapsulation with the ESP protocol	-13%	-22%	-28%	-19%	-23%

Table 3.7. *Summary of the impact of securing on throughput, RTT and number of packets processed per second*

3.3.3.3. *Availability of the router over one week*

Another element we would like to study is the "long-term" availability of the SNG router. We therefore focused on the behavior of the SNG router over a period of one week. In practice, large commercial carriers are rebooted each morning for the first flight of the day, but the aircraft can in exceptional circumstances function for up to 100 h during long-haul flights. The router is therefore used for less than one week, and hence our choice.

We therefore implemented the network topology with two hosts and two routers shown in Figure 3.11. It was configured to activate the security services (encryption and signature) and a load traffic of 20 Mbps in UDP was generated, deliberately lower than the maximum traffic of 22 Mbps that the SNG router can reach with these parameters. Loss was therefore studied during this 168 h experiment.

Out of the 1,004,155,245 packets sent by H1, 2,685 were not received by H2. The communication integrity rate is therefore higher than 99,999% (1 – 2.67e-6), i.e. the router met the most restrictive requirements for critical aeronautical communications specified by international norms (see [COC 07]). The router operated non-stop over the course of the week, i.e. it had perfect availability.

Our methodology suggests the use of an IMA compatible system, therefore for a system equipped with mechanisms that detect software and hardware faults and can correct them (by restarting the system, stopping it or by modifying its configuration), mechanisms such as *Health Manager* and *watchdog*, which are integrated into the Sysgo PikeOS operating system used for our SNG routers. For the SNG router, these mechanisms were configured to automatically restart in the event of a breakdown.

We tried to identify the source of the lost packets we observed. However, this loss appears to be random with regard to time: the time between two losses cannot be predicted.

The suspected sources are the two SNG routers (their software and hardware), the category 6 three meter long RJ45 cables and the H1 and H2 hosts (Linux Debian systems). Measuring H1 and H2 enabled us to determine that the loss rate for maximum size UDP packets by these end hosts was 1e-8.

To establish an increase in the loss rate for packets by an SNG router, we hypothesized that the connections made by the RJ45 cable were perfect. This enabled us to establish that the loss rate for packets by one SNG router is at most 1.34e-6.

This indicative value must, however, be put in perspective: it was calculated for a topology with one path and three subnetworks that only transports data with UDPv6 datagrams, the packets still had an MTU size of 1,500 bytes and the calculation load required by SNG routers was close to 90% of its previously measured maximum capacities.

3.3.3.4. *Overall performance of the system*

The overall performance of the SNG system is summarized in Figure 3.15. The two main parameters we will discuss in this section are the minimum overall processing capacity the router offers and the resulting maximum processing time for each packet it sends.

The left-hand side of Figure 3.15 illustrates that the minimum overall processing capacity of the SNG router is about 48 Mbps. This is enough throughput to enable the traffic generated by the intermediary of technologies currently available in the aeronautical industry to be transmitted, transferring information from the aircraft to the ground. This means that our system is capable of supporting the characteristics of all current aeronautical communication systems as well as the future innovations with regard to the interconnection between the aircraft and the different ground services available.

Moreover, by analyzing the maximum processing time metric, it can be seen (right-hand side of Figure 3.15) that the SNG router gives a time that is much lower than all the air–ground communication systems with which it could be interfaced. Indeed, the SNG router gives a maximum processing time of 1 ms whereas recommendations for air–ground communication specify a minimum transportation time of (between 100 ms and 1 s) for the majority of aeronautical services (see [COC 07] for more details).

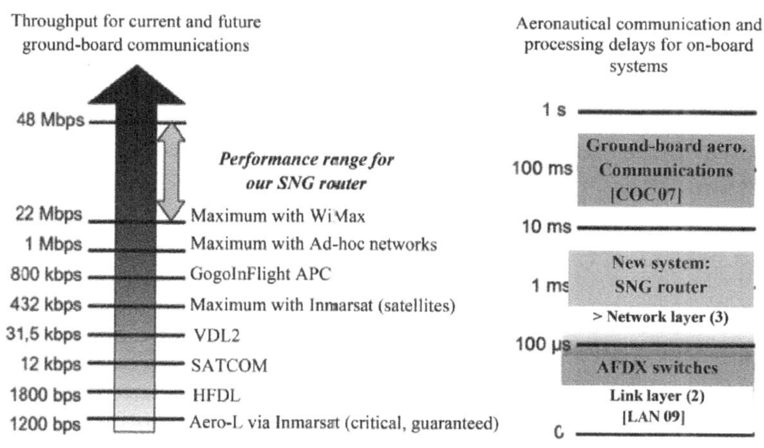

Figure 3.15. *Overall performance of the SNG system*

To conclude these evaluation results for the performance of the SNG router, it can be stated that the model-based rapid prototyping methodology presented in this book reached the performance objectives sought by current and future onboard aeronautical systems. Furthermore, as mentioned in the previous section, the improved performance of the final system is accompanied by a shortening of the time required to develop the system, in comparison to traditional software engineering. This time is gained from the use of verification methods derived from models and from the integration of certified tools which enable source code to be autogenerated for the final system. Using these tools also liberates developers from the large number of long and tiresome unit tests, enabling them to favor formal verification methods.

3.4. Chapter summary

This chapter presented several performance measures for the SNG router, after having listed the metrics of interest for the evaluation of the router and detailed the network topologies we used. Far from being exhaustive due to the large number of parameters likely to influence the performance of our onboard SNG router, we attempted to cover a maximum number of possibilities to validate the functionalities and more importantly to enable significant trends to be extracted from our results, i.e. the impact of these parameters on different measurable metrics.

However, the majority of these measurements are specific to our version of the software that is derived from our modeling of functionalities described at the beginning of this chapter. These measurements are also specific to our experiment hardware. As such, a change in hardware and/or models could strongly impact the raw values of these measurements. Trends should, however, remain; for instance, TCP throughputs should remain lower than UDP throughputs and securing should always reduce performance.

The performance of our system is widely acceptable with regard to industrial aeronautical restrictions. As such, from our point of view, our product, the SNG router, is ready to enter the next industrial phase for product development and leave the prototyping phase. It has been proved that the methodology presented in Chapter 2 and applied in this chapter enables a complete and functional system to be rapidly realized.

Conclusion

Here we will review the different stages discussed in this book: the history of model-based approaches, the presentation of an original rapid prototyping methodology for avionic software and an illustration of this methodology via the presentation of the secure next generation (SNG) router. The various applications that could use this methodology in the future will also be discussed. Moreover, some scientific approaches that are currently being investigated to improve the system verification phase will be described. For instance, research into validation based on a qualitative and quantitative analysis of models aimed at revealing the properties from the beginning of the design phase, thus shortening the software source code testing phase (in comparison to more classic approaches to design).

Chapter 1 of this book focused on the history of design methods based on models and the Unified Modeling Language (UML). Over the last decade, the latter has become the essential language for software development, whether it is generalist or, more recently, aeronautical. Its expressive power enables it to be used in any project in the world of software development, which until now has been characterized by the large diversity of the methods and the languages proposed.

Avionic systems are critical: if they fail, they put the lives of the humans onboard the airplane at risk. Aircraft designers are, therefore, required to guarantee the innocuity of the systems, to certify these systems as safe. A large number of standards and norms are used in aeronautics, a domain in which modern systems rub shoulders with older systems; the different formal verification methods for software are, nowadays, mature and aeronautical manufacturers use them increasingly to automate the validation of their products. The recent update of the DO178B means that we can serenely imagine using these methods for new development projects going forward.

The countless restrictions related to the development of critical avionic software forced us to think carefully about which process to adopt for the development of an innovative "router" system. This led us to create a development and prototyping methodology, spanning from the initial phase of writing system requirements to the acceptance test, i.e. the verification and the validation of the system on an emulator or real system.

Using the power of virtualization, model-based development, the safety provided by the Integrated Modular Avionics (IMA) and the security that comes alongside Multi Independent Level of Security (MILS) concepts, the methodology guides development teams toward experiments, speeding up many tasks by automating the stages while taking into consideration the evaluation and certification restrictions for the final system produced, restrictions that are commonly required in the context of critical onboard software products.

In order for the methodology to be used for a large number of applications, and not only for our software that is part of a computer network system, the seven stages of the methodology are flexible. For instance, designers can choose countless different tools for the realization of each stage as well as the variations necessary to adapt the methodology to the specificities of each software development project.

Other features of the methodology complete the advantages accompanying its application. For this reason, the concepts of partition classes and partition instances were created to facilitate the development of the software architecture as well as the reuse of modeling in order to rapidly generate variants of the software.

Having carefully thought about the development methodology, in Chapter 3 we applied it to an avionic Internet Protocol (IP) router and explained the specificities of the aeronautical environment and the tough security and safety restrictions that apply.

We, therefore, created models for the different network protocols that are currently in use, or being planned, for aeronautics: IPv4 and IPv6 that delve further into the description of the implementation of packet routing and filtering. High-level models are thus used to generate the source code to implement protocols so that they can be formally validated.

Implementing protocols extends to modeling using protocol state machines until they are tested and used in a real system. Implementing protocols spans from the "high" layers of functionalities (such as the routing and filtering mechanisms) to the "low" physical layers of data exchange between systems via electronic signals, while taking into account the specificities of the avionic context (such as the absence of the ARP in the Avionics Full-Duplex Switched Ethernet (AFDX)).

In addition, our SNG router has strongly proved the applicability of our methodology: the latter came about through ideas and concepts and has been validated, enabling a complete and functional SNG router to be developed, complete with data security mechanisms provided with the information necessary for its security evaluation and safety certification.

The reflection period for the development process of the SNG router (Chapter 2) and its implementation (Chapter 3) are completed by evaluating the performance of our SNG router.

For the performance evaluation phase, we studied the network traffic for aeronautics and developed a generic software tool that generates *ad hoc* network traffic to create a realistic load on the network, corresponding to our projections for avionic and aeronautic communications for 2020. We then conducted experiments following different network topologies and measured different performance criteria for our SNG router.

The results of our performance evaluation are positive: the throughput, delay, jitter and availability measurements are acceptable with regard to the existing restrictions, as well as those predicted for the near future in the aeronautical domain, regardless of the transport protocol tested (Transmission Control Protocol (TCP) and User Datagram Protocol (UDP)), the network flows tested (the impacts were studied for 1–16 simultaneous flows) and the size of the packets exchanged (limited to 60 bytes and an Maximum Transmission Unit (MTU) of 1,500 bytes).

The performance of our router, in addition to the development methodology that enables our router software to be rapidly produced or reproduced, therefore leads us to conclude that the router is close to moving into the next manufacturing stage, Technical Readiness Level (TRL) 5. In this stage, the router is tested in an environment simulating real avionic architecture, including, for instance, AFDX flows. This stage is, in fact, an intermediate stage between where we are now and TRL 8, where the final system is used in an operational situation, the final objective for our system.

This research enabled us to explore IMA and MILS concepts in more depth. Our development methodology approach uses these two concepts simultaneously; these concepts are, however, attached to two sets of standards: one for IMA and the other for MILS, which are used by two different communities (the aeronautical community and the military community). These concepts are, however, partly redundant: although they have different objectives, security for MILS and safety for IMA, the mechanisms used to realize them are similar. A significant evolution would, therefore, be to combine these two concepts and to formalize their unification with a standard.

Our methodology (presented in Chapter 2) is open to the use of new tools and new intermediate languages. We had, in fact, thought about using a "functional programming tool" chain based on the CaML language, but we did not explore this approach in detail due to the requirements of our SNG router. A functional chain could, however, prove to be much better adapted than our "imperative programming" chain for research projects based on pure mathematics and deriving from the domain of operations research.

Our methodology splits the partition classes from the partition instances; similarly, an object class is split from an object instance in *model-based design*. This division was intended, in particular, to prepare for the next mutation of architectures based on monocore processors and to manage the threads of execution sequentially for multicore or multiprocessor, and even massively multicore (General-Purpose Computation on Graphics Processing Unit (GPGPU)) – where each instance can be associated with its core – architectures of the future. Multicore systems have demonstrated their effectiveness in different domains; however, the world of aeronautics is still hesitant to integrate them due to functional security considerations. Avionic software development projects that have followed our methodology should be easy to adapt and future parallel architectures should cost less, given that GPGPUs will be available on the aeronautical hardware platform market. This will be possible as working through all the software development and validation processes will be avoided. We finally tested the methodology on the SNG router with the C language. Although we had tools to use the Ada language as an intermediate language, we were satisfied with evaluating performance on a version generated in C. However, the Ada language potentially has many advantages in terms of performance and optimization, as well as in terms of the security and safety of software function. Additional research would be beneficial to evaluate the impact of the intermediate language on the application of the methodology.

The methodology enables additional partition classes dedicated to complex tasks to be integrated. It is, therefore, possible to add the functionality of *Deep Packet Inspection* to the SNG router, which filters the non-critical flows more effectively. These advanced filtering functionalities are currently being studied by aircraft manufacturers and their equipment suppliers, but their integration has slowed due to security and safety restrictions; the methodology facilitates this integration as it guarantees high levels of innocuity and immunity, thus enabling us to imagine their integration in the short term.

Using the methodology for the router enabled certain router functionalities, as well as certain properties, to be formally validated: absence of blocking, absence of surpassing the capacity of the scalar of the programming languages, etc. Realized at the initial modeling phase, this validation avoids certain bugs being debugged at a late stage, the cost of which to correct or overcome is high, in comparison to the cost of implementing formal methods on models.

This book has shown that it is possible to rapidly produce quality software that meets the tough certification and evaluation restrictions by using a set of carefully chosen tools. Moreover, aircraft manufacturers can henceforth realize an avionic architecture with not only the existing network equipment, the physical layer (1) and the link layer (2), but also using the opportunities offered by the network layer (3) via the aeronautical SNG router. Finally, the path for modeling aeronautical software was strengthened by using the same model twice: the formal validation of functionalities and the automatic generation of its implementation in operational systems.

In addition to the research we conducted into the methodological and technical development of new onboard avionic equipment, our SNG router, the ideas resulting from this research open the door to new research, creating new opportunities and improving the effectiveness of aeronautical systems, while maintaining high levels of air travel security and safety.

Bibliography

[ABR 96] ABRIAL J.-R., *The B-Book: Assigning Programs to Meanings*, Cambridge University Press, Cambridge, 1996.

[ADA 05] ADAMS M.M., CLAYTON P.B., TUDOR N.J., "CLAWZ: cost-effective formal verification for control systems", *26th DASC*, pp. 10.D.6-1–10.D.6-10, 2005.

[ADA 13] ADACORE, GNAT Pro, Ada compiler by AdaCore, 2013. Available at http://www.adacore.com/.

[AGR 11] AGROU H., SAINRAT P., GATTI M., *et al.*, "A design approach for predictable and efficient multi-core processor for avionics", *IEEE/AIAA 30th Digital Avionics Systems Conference*, pp. 7D3-1–7D3-11, October 2011.

[AOU 07] AOUN C., DAVIES E., "Reasons to move the network address translator – protocol translator (NAT-PT) to historic status", *RFC 4966 (Informational)*, July 2007.

[ARI 01] AIRLINES ELECTRONIC ENGINEERING COMMITTEE (AEEC) ARINC622-4, ARINC Report 622-4, ATS Data Link Applications over ACARS Air-Ground Network, October 2001.

[ARI 05a] AIRLINES ELECTRONIC ENGINEERING COMMITTEE (AEEC) ARINC623-3, ARINC Report 623-3, Character-Oriented Air Traffic Service (ATS) Applications, April 2005.

[ARI 05b] AIRLINES ELECTRONIC ENGINEERING COMMITTEE (AEEC) ARINC664P5, ARINC Report 664P5, Aircraft Data Network, Part 5, Network Domain Characteristics and Functional Elements, April 2005.

[ARI 05c] AIRLINES ELECTRONIC ENGINEERING COMMITTEE (AEEC) ARINC811, ARINC Standards 811, Commercial Aircraft Information Security Concepts of Operation and Process Framework, December 2005.

[ARI 06a] AIRLINES ELECTRONIC ENGINEERING COMMITTEE (AEEC) ARINC618-6, ARINC Report 618-6, Air/Ground Character-Oriented Protocol Specification, June 2006.

[ARI 06b] AIRLINES ELECTRONIC ENGINEERING COMMITTEE (AEEC) ARINC653P3, ARINC Report 653P3, Avionics Application Software Standard Interface, Part 3 – Conformity Test Specification, October 2006.

[ARI 08] AIRLINES ELECTRONIC ENGINEERING COMMITTEE (AEEC) ARINC821, ARINC Standards 821, Aircraft Network Server System (NSS) Functional Definition, December 2008.

[ARI 09] AIRLINES ELECTRONIC ENGINEERING COMMITTEE (AEEC) ARINC619-3, ARINC Report 619-3, ACARS Protocols for Avionic End Systems, June 2009.

[ARI 10] AIRLINES ELECTRONIC ENGINEERING COMMITTEE (AEEC) ARINC653P1-3, ARINC Report 653P1-3, Avionics Application Software Standard Interface, Part 1 – Required Services, November 2010.

[ARI 12a] AIRLINES ELECTRONIC ENGINEERING COMMITTEE (AEEC) ARINC620-7, ARINC Report 620-7, Data Link Ground System Standard and Interface Specification (DGSS/IS), June 2012.

[ARI 12b] AIRLINES ELECTRONIC ENGINEERING COMMITTEE (AEEC) ARINC653P2-2, ARINC Report 653P2-2, Avionics Application Software Standard Interface, Part 2 – Extended Services, June 2012.

[ARI 12c] AIRLINES ELECTRONIC ENGINEERING COMMITTEE (AEEC) ARINC653P4, ARINC Report 653P4, Avionics Application Software Standard Interface, Part 4 – Subset Services, June 2012.

[AUD 09] AUDIBERT L., *UML 2: de l'apprentissage à la pratique*, Ellipses Marketing, Paris, 2009.

[BAH 03] BAHILL T., DANIELS J., "Using objected-oriented and UML tools for hardware design: a case study", *Systems Engineering*, vol.6, no. 1, pp.28–48, 2003.

[BAU 09] BAUDIN P., CUO P., FILLIÂTRE J.-C., *et al.*, "ACSL: ANSI/ISO C specification language", *CEA LIST and INRIA*, version 1.7, 2009–2013. Available at http://frama-c.com/download/acsl.pdf.

[BEN 12] BEN MAHMOUD M.S., Addressing security challenges in emerging data-based aeronautical communications, PhD Thesis, National Institute of Applied Sciences in Toulouse (INSA Toulouse), February 2012.

[BER 14] BERTOT Y., CASTÉRAN P., *Le Coq'Art (v8)*, January 2014. Available at http://www.labri.fr/perso/casteran/CoqArt/coqartF.pdf.

[BES 13] BESSE F., Réseaux ad hoc aéronautiques, PhD Thesis, French Civil Aviation University (ENAC), Toulouse, February 2013.

[BHA 05] BHATT D., HALL B., DAJANI-BROWN S., *et al.*, "Model-based development and the implications to design assurance and certification", *24th Digital Avionics Systems Conference (DASC '05)*, pp. 10.D.3-1–10.D.3-13, 2005.

[BOO 93] BOOCH G., *Object-Oriented Analysis and Design with Applications*, 2nd. ed., Benjamin Cummings, Redwood City, 1993.

[BOO 08] BOOCH G., JACOBSON I., RUMBAUGH J., *OMG Unified Modeling Language Specification*, version 1.3, 1st ed., March 2000, reedition 12 August 2008.

[BRI 96] BRIFFAULT X., SABAH G., *Smalltalk: programmation orientée objet et développement d'applications*, Eyrolles, Paris, 1996.

[BUS 04] BUSSER R.D., BLACKBURN M.R., NAUMAN A.M., *et al.*, "Reducing cost of high integrity systems through model-based testing", *23rd Digital Avionics Systems Conference (DASC '04)*, vol. 2, pp. 6.B.1–6.B-13, 2004.

[CC 09a] CC1, Common Criteria for Information Technology Security Evaluation, Part 1: Introduction and General Model, July 2009.

[CC 09b] CC2, Common Criteria for Information Technology Security Evaluation, Part 2: Security Functional Components, July 2009.

[CC 09c] CC3, Common Criteria for Information Technology Security Evaluation, Part 3: Security Assurance Components, July 2009.

[CC 09d] CC4, Common Methodology for Information Technology Security Evaluation. Part 4: Evaluation Methodology, July 2009.

[CLA 96] CLARKE E.M., WING J.M., *et al.*, "Formal methods: state of the art and future directions", *ACM Computing Surveys*, vol. 28, p. 4, December 1996.

[COC 07] COCR, Communications Operating Concept and Requirements for Future Radio System, version 2.0, Eurocontrol, 1 January 2007.

[CLE 13] CLEARSY, *Atelier-B*, April 2013. Available at http://www.atelierb.eu/.

[CNE 09] CNES, "SATCOM for ATM: estimation of capacity required for AMS(R)S communications around 2020 over European area", July 2009.

[COM 13a] COMBS G., Wireshark, Go Deep, 2013. Available at http://www.wireshark.org/.

[COM 13b] COMBS G., Dumpcap – dump network traffic, June 2013. Available at http://www.wireshark.org/docs/man-pages/dumpcap.html.

[CON 06] CONTA A., DEERING S., GUPTA M., Internet Control Message Protocol (ICMPv6) for the Internet Protocol Version 6 (IPv6) Specification, RFC 4443 (Draft Standard), updated by RFC 4884, March 2006.

[DA 11] DA SILVA G.C., CORREA S., DA CUNHA A.M., *et al.*, "A comparison between automated generated code tools using model based development", *IEEE/AIAA 30th Digital Avionics Systems Conference*, pp. 7E4-1–7E4-9, October 2011.

[DEE 95] DEERING S., HINDEN R., Internet Protocol, Version 6 (IPv6) Specification, RFC 1883 (Proposed Standard), made obsolete by RFC 2460, December 1995.

[DEE 98] DEERING S., HINDEN R., Internet Protocol, Version 6 (IPv6) Specification, RFC 2460 (Draft Standard), updated by RFCs 5095, 5722, 5871, 6437, 6564, December 1998.

[DES 94] DESFRAY P., *Object Engineering: The Fourth Dimension*, Addison-Wesley, Indianapolis, 1994.

[DES 10] DESPRES R., "IPv6 rapid deployment on IPv4 infrastructures", *6th RFC 5569 (Informational)*, January 2010.

[DIA 13] DIA, Dia, éditeur libre de diagrammes, April 2013. Available at http://live.gnome.org/Dia/.

[DO 76] DO-163, SC-131 DO-163, Minimum performance standards-airborne HF radio communications transmitting and receiving equipment operating within the radio-frequency range of 1.5 to 30 megahertz, Radio Technical Commission for Aeronautics, 19 March 1976.

[DO 79] DO-169, SC-140 DO-169, VHF air-ground communication technology and spectrum utilization, Radio Technical Commission for Aeronautics, 20 July 1979.

[DO 92] DO-178B, SC-167 DO-178B, Software considerations in airborne systems and equipment certification, Radio Technical Commission for Aeronautics, 1 December 1992.

[DO 00] DO-265, SC-188 DO-265, Minimum operational performance standards for aeronautical mobile high frequency data link (HFDL). Radio Technical Commission for Aeronautics, 14 December 2000.

[DO 05a] DO-281A, SC-172 DO-281A, Minimum operational performance standards for aircraft VDL mode 2 physical, link and network layer, Radio Technical Commission for Aeronautics, 8 November 2005.

[DO 05b] DO-297, SC-200 DO-297, Integrated modular avionics (IMA) development guidance and certification considerations, Radio Technical Commission for Aeronautics, 8 November 2005.

[DO 11] DO-178C, SC-205 DO-178C, Software considerations in airbone systems and equipment certification, Radio Technical Commission for Aeronautics, 13 December 2011.

[DO 12a] DO-281B, SC-214 DO-281B, Minimum operational performance standards (MOPS) for aircraft VDL mode 2 physical link and network layer, Radio Technical Commission for Aeronautics, 21 March 2012.

[DO 12b] DO-331, SC-205 DO-331, Model-based development and verification supplement to DO-178C and DO-278A, Radio Technical Commission for Aeronautics, 5 January 2012.

[DO 12c] DO-332, SC-205 DO-332, Object-oriented technology and related techniques supplement to DO-178C and DO-278A, Radio Technical Commission for Aeronautics, 5 January 2012.

[DO 12d] DO-333, SC-205 DO-333, Formal methods supplement to DO-178C and DO-278A, Radio Technical Commission for Aeronautics, 5 January 2012.

[DOV 02] DOVROLIS C., JAIN M., "Pathload, a measurement tool for end-to-end available bandwidth", *Passive Active Measurements Conference (PAM)*, January 2002.

[DOV 06] DOVROLIS C., JAIN M., Pathload, a measurement tool for the available bandwidth of network paths, 2006. Available at http://www.cc.gatech.edu/fac/Constantinos.Dovrolis/bwest/pathload.html.

[DUB 03] DUBRAWSKY I., "Firewall evolution – deep packet inspection", *Security Focus*, vol. 7, no. 1716, p. 16, August 2003.

[DUB 12] DUBOIS H., IBANEZ V., LOPEZ C., *et al.*, "The product line engineering approach in a model-driven process", *European Congress on Embedded Real-Time Software (ERTS)*, 2012.

[ECL 13] ECLIPSE, An open development platform comprised of extensible frameworks, tools and runtimes for building, April 2013. Available at http://www.eclipse.org/.

[ESA 06] ESA WG HOOD, European Space Agency – Software Engineering and Standardisation Workgroup, The HOOD method, 2006. Available at http://www.esa.int/TEC/Software_engineering_and_standardisation/TECKLAUXBQE_0.html.

[FRA 14] FRAMAC, CEA-LIST 2007-2013, The Frama-C official website, January 2014. Available at http://www.frama-c.com.

[GAL 11] GALLET B., BRINTON C., "Massively parallel processing for dynamic airspace configuration", *IEEE/AIAA 30th Digital Avionics Systems Conference*, pp. 3A5-1–3A5-9, October 2011.

[GAR 09] GARSIDE R., PIGHETTI F., "Integrating modular avionics: a new role emerges", *IEEE Aerospace and Electronic Systems Magazine*, vol. 24, no. 3, pp. 31–34, March 2009.

[GAT 13] GATES M., FERGUSON J., DUGAN J., *et al.*, Iperf, a modern alternative for measuring TCP and UDP bandwidth performance, June 2013. Available at http://iperf.fr/.

[GCC 13] GCC, The GNU C Compiler, April 2013. Available at http://gcc.gnu.org/.

[GHA 10] GHAFARI N., "Applying formal methods to airborne software", *29th DASC Proceedings*, October 2010.

[GIG 12] GIGANTE G., PASCARELLA D., "Formal methods in avionic software certification: the DO-178C perspective", *Leveraging Applications of Formal Methods, Verification and Validation, Applications and Case Studies,* Lecture Notes in Computer Science, vol. 7610, pp. 205–215, 2012.

[HAL 77] HALSTEAD M.H., "Elements of software science", *Operating and Programming Systems Series 2*, Elsevier, New York, 1977.

[HOA 69] HOARE C.A.R., "An axiomatic basis for computer programming", *Communications of the ACM*, vol. 12, no. 10, pp. 576–585, October 1969.

[HOR 10] HORVÁTH A., VARRÓ D., SCHOOFS T., "Model-driven development of ARINC 653 configuration tables", *IEEE/AIAA 29th Digital Avionics Systems Conference (DASC)*, 2010.

[HUE 85] HUET G., SUÁREZ A., WEIS P., *et al.*, The Categorical Abstract Machine Language (CaML), 1985.

[HUI 06] HUITEMA C., Teredo: Tunneling IPv6 over UDP through Network Address Translations (NATs), RFC 4380 (Proposed Standard), updated by RFCs 5991, 6081, February 2006.

[IB 10] IB, TONU N.K., Gene-Auto: Automatic Software Code Generation for Real-Time Embedded Systems: D2.55 Gene-Autoinstallation and Usage Manual, June 2010.

[IBM 13] IBM, IBM Rational DOORS, April 2013. Available at http://www-01.ibm.com/software/awdtools/doors/productline/.

[ICH 83] ICHBIAH J., The Ada Language, 1983. Available at http://www.adaic.com/.

[INM 13] INMARCHAT, The mobile satellite company(TM), June 2013. Available at http://www.inmarchat.com/.

[INR 13] INRIA, Metalau Project INRIA, Scicos, simulateur graphique de systèmes dynamiques, April 2013. Available at http://www.scicos.org/.

[IRI 13] IRIDIUM, Site relatif à la constellation "Iridium" de satellites, pour l'aéronautique, June 2013. Available at http://www.iridium.com/IridiumConnected/IridiumAtWork/-Aviation.aspx.

[ISO 01] ISO/IEC 19501:2005, Information Technology – Open Distributed Processing – Unified Modeling Language (UML), Version 1.4.2, 2001. Available at http://www.omg.org/spec/UML/ISO/19501/PDF/.

[IZE 10] IZERROUKEN N., THIRIOUX X., PANTEL M., *et al.*, "Design and development of an embedded aeronautical, certifying an automated code generator using formal tools: preliminary experiments in the gene-auto project", *European Congress on Embedded Real-Time Software (ERTS)*, Toulouse, 2010.

[JAC 04] JACOBS R.W., "Model-driven development of command and control capabilities for joint and coalition warfare", *Command and Control Research and Technology Symposium*, 15–17 June 2004.

[JAC 92] JACOBSON I., CHRISTERSON M., JONSSON P., *et al.*, *Object Oriented Software Engineering: A Use Case Driven Approach*, Addison Wesley/ACM Press, New York, 1992.

[JAC 98] JACOBSON I., BOOCH G., RUMBAUGH J., *The Unified Software Development Process*, Addison Wesley Longman, Chicago, 1998.

[JAC 00] JACOBSON I., BOOCH G., RUMBAUGH J., *Le Guide de l'utilisateur UML*, Eyrolles, Paris, 2000.

[JAE 10] JAEGER E., Study of the benefits of using deductive formal methods for secure developments, PhD Thesis, LIP6, March 2010.

[JOB 11] JOBREDEAUX R., WANG T.E., FERON E.M., "Autocoding control software with proofs I: annotation translation", *IEEE/AIAA 30th Digital Avionics Systems Conference*, pp. 7C1-1–7C1-13, October 2011.

[JOY 10] JOYCE J., "Formal methods in RTCA, DO-178C", *29th DASC Proceedings*, October 2010.

[KOC 96] KOCHER P.C., *Timing Attacks on Implementations of Diffie-Hellman, RSA, DSS and Other Systems*, Springer, Berlin, vol. 1109, 1996.

[LAN 09] LAND I., ELLIOTT J., Architecting ARINC 664, partie 7 (AFDX) Solutions, Xilinx, May 2009.

[LEL 94] LELAND W.E., TAQQU S.M., WILSON D.V., "On the self-similar nature of ethernet traffic", *IEEE/ACM Transactions on Networking*, vol. 2, no. 1, pp. 203–213, February 1994.

[LER 06] LEROY X., "Formal certification of a compiler back-end", *ACM CE, ACM*, January 2006.

[LIB 13] LIBREOFFICE, Communauté LibreOffice francophone, LibreOffice, April 2013. Available at http://fr.libreoffice.org/.

[MAT 11] MATOLAK D.W., WU Q., SANCHEZ-SANCHEZ J.J., *et al.*, "LTE performance in the airport surface area channel", *IEEE/AIAA 30th Digital Avionics Systems Conference*, pp. 4D3-1–4D3-7, October 2011.

[MAT 13a] MATHWORKS, The MathWorks, Matlab R2012b, April 2013. Available at http://www.mathworks.fr/.

[MAT 13b] MATHWORKS, The MathWorks, Simulink, logiciel de modélisation système multiphysique, April 2013. Available at http://www.mathworks.com/products/simulink/.

[MAT 13c] MATHWORKS, The MathWorks, Stateflow, logiciel de modélisation de diagrammes à états, April 2013. Available at http://www.mathworks.com/products/stateflow?s_cid=wiki_stateflow_2.

[MIL 03] MILLER R., "Designing web services using UML", *BorCon Conference, Agile Processes Workshop*, 2003.

[MIS 98] MISRA-C, Motor Industry Software Reliability Association, 1998. Available at http://www.embedded.com/columns/beginers corner/9900659 ?_requestid=427335.

[MOY 08] MOY Y., "Sufficient preconditions for modular assertion checking", *9th International Conference Proceedings (VMCAI '08)*, San Francisco, vol. 4905, pp. 188–202, 7–9 January 2008.

[OAC 07] OACI, Annexe 10 de la Convention relative à l'aviation civile internationale, Télécommunications aéronautiques, vol. III, Communication Systems (part I – Digital Data Communication Systems, part II – Voice Communication Systems), 2nd ed., July 2007. Available at http://www.bazl.admin.ch/dokumentation/grundlagen/02643/index.html?lang=fr&download=NHzLpZeg7t,lnp6I0NTU042l2Z6ln1ae2IZn4Z2qZpnO2Yuq2Z6gpJCDeoB9fGym162epYbg2c_JjKbNoKSn6A.

[OMA 05] OMAN P., ALVES-FOSS J., HARRISON W.S., *et al.*, "The MILS architecture for high-assurance embedded systems", *International Journal of Embedded Systems*, vol. 1, pp. 1–9, 2005.

[PIK 11] PIKEOS, Sysgo PikeOS Real-Time Operating System, December 2011. Available at http://www.sysgo.com/products/pikeos-rtos-and-virtualization-concept/.

[POL 11] POLGAR B., "Model-based integration, execution and certification of development tool-chains", *Formal Methods for Automation and Safety in Railway and Automotive Systems*, Department of Measurement and Information Systems, Budapest University of Technology and Economics, Hungary, pp. 227–235, 2011.

[POO 87] POOLEY R., *An Introduction to Programming in Simula*, Alfred Waller Ltd., Bedford, 1987.

[POS 80] POSTEL J., User Datagram Protocol, RFC 768, August 1980.

[POS 81a] POSTEL J., Internet Protocol, RFC 791, updated by RFCs 1349, 2474, September 1981.

[POS 81b] POSTEL J., Transmission Control Protocol, RFC 793, updated by RFCs 1122, 3168, 6093, 6528, September 1981.

[POS 81c] POSTEL J., Internet Control Message Protocol, RFC 792, updated by RFCs 950, 4884, 6633, September 1981.

[QEM 11] QEMU, A generic and open source machine emulator and virtualizer, December 2011. Available at http://wiki.qemu.org/.

[RAY 01] RAYADURGAM S., HEIMDAHL M.P.E., "Coverage based test-case generation using model checkers", *ECBS*, pp. 83–91, 2001. Available at: http://ieeexplore.ieee.org/xpl/articleDetails.jsp?arnumber=922409.

[RHE 08] RHEE I., HA S., XU L., "CUBIC: a new TCP-friendly high-speed TCP variant", *ACM SIGOPS Operating System Review*, vol. 42, pp. 64–74, July 2008.

[RIE 11] RIEGER G., Multipurpose relay (SOcket CAT), relay for bidirectional data transfer between two independent data channels, December 2011. Available at http://stackoverflow.com/tags/socat/info.

[RIT 78] RITCHIE D., KERNIGHAN B., *The C Programming Language*, 1st ed., Prentice Hall Software Series, New Jersey, 1978.

[ROQ 06] ROQUES P., *UML 2 par la pratique – Etudes de cas et exercices corrigés*, Eyrolles, Paris, 2006.

[RUG 08] RUGINA A.E, THOMAS D., OLIVE X., *et al.*, "Gene-Auto: Automatic Software Code Generation for Real-Time Embedded Systems", S. ASTRIUM Satellites SA and Thales A. Space, Dasia, 2008.

[RUG 10] RUGINA A.E., DALBIN J.C., "Experiences with the gene-auto code generator in the aerospace industry", *Astrium + Airbus*, 2010.

[RUM 90] RUMBAUGH J., BLAHA M., PREMERLANI W., *et al.*, *Object-Oriented Modeling and Design*, Prentice Hall, New Jersey, 1990.

[RUS 81] RUSHBY J.M., "Design and verification of secure systems", *SIGOPS Operation System Review*, vol. 15, no. 5, pp. 12–21, December 1981.

[SCA 13] SCADE, Software Critical Application Development Environment (SCADE), a lustre-based IDE generating C-code, April 2013. Available at http://www.estereltechnologies.com/-products/scade-system/.

[STR 00] STROUSTRUP B., *The C++ Programming Language*, special ed., Addison Wesley, New York, 2000.

[SUB 03] SUBBIAH S., NAGARAJ S., "Issues with object orientation in verifying safety-critical systems", *6th IEEE International Symposium on Object-Oriented Real-Time Distributed Computing*, pp. 99–104, 2003.

[SYS 10] SYSGO, PikeOS, Safe and Secure Virtualization", 2010. Available at http://www.sysgo.com/products/pikeos-rtos-and-virtualization-concept/.

[SYS 12] SYSGO, CODEO, Eclipse based IDE, 2012. Available at http://www.sysgo.com/products/pikeos-rtos-and-virtualizationconcept/eclipse-based-codeo/.

[TAR 83] TARDIEU H., *La méthode MERISE: principes et outils*, Editions d'Organisation, Paris, 1983.

[TOO 08] TOOM A., *et al.*, "Gene-auto: an automatic code generator for a safe subset of simulink/stateflow and scicos", *4th European Congress ERTS Embedded Real Time Software*, 2008.

[TOO 10] TOOM A., IZERROUKEN N., NAKS T., *et al.*, "Towards reliable code generation with an open tool: evolutions of the gene-auto toolset", *5th European Congress ERTS Embedded Real Time Software*, 2010.

[VAR 12] VARET A., LARRIEU N., MACABIAU C., "Design and development of an embedded aeronautical router with security capabilities", *Integrated Communication, Navigation and Surveillance Conference*, Washington, Colombia, 2012.

[UCH 09a] UCHENICK G.M., "Introduction to security for integrated modular avionics", *28th DASC Proceedings*, October 2009.

[UCH 09b] UCHENICK G.M., "Multiple independant levels of safety and security: high assurance architecture for integrated modular systems", *28th DASC Proceedings*, October 2009.

[UIT 13] UIT, Engagée à connecter le monde – site *web* officiel, June 2013. Available at http://www.itu.int/fr/.

[VAN 13] VAN JACOBSON C.L., MCCANNE S., tcpdump – dump traffic on a network, June 2013. Available at http://www.tcpdump.org/.

[VOU 96] VOUILLON J., DOLIGEZ D., RÉMY D., *et al.*, Objective Caml, INRIA, 1996. Available at http://caml.inria.fr/.

[VXW 11] VXWORKS, WindRiver VxWorks Real-Time Operating System, December 2011. Available at http://www.windriver.com/products/vxworks/.

[WIE 12] WIELS V., DELMAS R., DOOSE D., *et al.*, "Formal verification of critical aerospace software", *AerospaceLab Journal*, no. 4, pp. 1–8, May 2012.

[WIL 97] WILLINGER W., TAQQU M.S., WILSON D.V., "Self-similarity through high-variability: statistical analysis of Ethernet LAN traffic at the source level", *IEEE/ACM Transactions on Networking*, vol. 5, no. 1, February 1997.

[WIN 09] WINDRIVER, The Wind River VxWorks RTOS, 2009. Available at http://www.windriver.com/products/vxworks/.

Index

Other titles from

ISTE

in

Computer Engineering

2014

BOULANGER Jean-Louis
Formal Methods Applied to Industrial Complex Systems

BOULANGER Jean-Louis
Formal Methods Applied to Complex Systems: Implementation of the B Method

GARDI Frédéric, BENOIST Thierry, DARLAY Julien, ESTELLON Bertrand, MEGEL Romain
Mathematical Programming Solver based on Local Search

KRICHEN Saoussen, CHAOUACHI Jouhaina
Graph-related Optimization and Decision Support Systems

OUSSALAH Mourad Chabane
Software Architecture 1

OUSSALAH Mourad Chabane
Software Architecture 2

QUESNEL Flavien
Scheduling of Large-scale Virtualized Infrastructures: Toward Cooperative Management

RIGO Michel
Formal Languages, Automata and Numeration Systems 1: Introduction to Combinatorics on Words
Formal Languages, Automata and Numeration Systems 2: Applications to Recognizability and Decidability

SAINT-DIZIER Patrick
Musical Rhetoric: Foundations and Annotation Schemes

TOUATI Sid, DE DINECHIN Benoit
Advanced Backend Optimization

2013

ANDRÉ Etienne, SOULAT Romain
The Inverse Method: Parametric Verification of Real-time Embedded Systems

BOULANGER Jean-Louis
Safety Management for Software-based Equipment

DELAHAYE Daniel, PUECHMOREL Stéphane
Modeling and Optimization of Air Traffic

FRANCOPOULO Gil
LMF — Lexical Markup Framework

GHÉDIRA Khaled
Constraint Satisfaction Problems

ROCHANGE Christine, UHRIG Sascha, SAINRAT Pascal
Time-Predictable Architectures

WAHBI Mohamed
Algorithms and Ordering Heuristics for Distributed Constraint Satisfaction Problems

ZELM Martin *et al.*
Enterprise Interoperability

2012

ARBOLEDA Hugo, ROYER Jean-Claude
Model-Driven and Software Product Line Engineering

BLANCHET Gérard, DUPOUY Bertrand
Computer Architecture

BOULANGER Jean-Louis
Industrial Use of Formal Methods: Formal Verification

BOULANGER Jean-Louis
Formal Method: Industrial Use from Model to the Code

CALVARY Gaëlle, DELOT Thierry, SÈDES Florence, TIGLI Jean-Yves
Computer Science and Ambient Intelligence

MAHOUT Vincent
Assembly Language Programming: ARM Cortex-M3 2.0: Organization, Innovation and Territory

MARLET Renaud
Program Specialization

SOTO Maria, SEVAUX Marc, ROSSI André, LAURENT Johann
Memory Allocation Problems in Embedded Systems: Optimization Methods

2011

BICHOT Charles-Edmond, SIARRY Patrick
Graph Partitioning

BOULANGER Jean-Louis
Static Analysis of Software: The Abstract Interpretation

CAFERRA Ricardo
Logic for Computer Science and Artificial Intelligence

HOMES Bernard
Fundamentals of Software Testing

KORDON Fabrice, HADDAD Serge, PAUTET Laurent, PETRUCCI Laure
Distributed Systems: Design and Algorithms

KORDON Fabrice, HADDAD Serge, PAUTET Laurent, PETRUCCI Laure
Models and Analysis in Distributed Systems

LORCA Xavier
Tree-based Graph Partitioning Constraint

TRUCHET Charlotte, ASSAYAG Gerard
Constraint Programming in Music

VICAT-BLANC PRIMET Pascale *et al.*
Computing Networks: From Cluster to Cloud Computing

2010

AUDIBERT Pierre
Mathematics for Informatics and Computer Science

BABAU Jean-Philippe *et al.*
Model Driven Engineering for Distributed Real-Time Embedded Systems 2009

BOULANGER Jean-Louis
Safety of Computer Architectures

MONMARCHE Nicolas *et al.*
Artificial Ants

PANETTO Hervé, BOUDJLIDA Nacer
Interoperability for Enterprise Software and Applications 2010

PASCHOS Vangelis Th
Combinatorial Optimization – 3-volume series
Concepts of Combinatorial Optimization – Volume 1
Problems and New Approaches – Volume 2
Applications of Combinatorial Optimization – Volume 3

SIGAUD Olivier *et al.*
Markov Decision Processes in Artificial Intelligence

SOLNON Christine
Ant Colony Optimization and Constraint Programming

AUBRUN Christophe, SIMON Daniel, SONG Ye-Qiong *et al.*
Co-design Approaches for Dependable Networked Control Systems

2009

FOURNIER Jean-Claude
Graph Theory and Applications

GUEDON Jeanpierre
The Mojette Transform / Theory and Applications

JARD Claude, ROUX Olivier
Communicating Embedded Systems / Software and Design

LECOUTRE Christophe
Constraint Networks / Targeting Simplicity for Techniques and Algorithms

2008

BANÂTRE Michel, MARRÓN Pedro José, OLLERO Hannibal, WOLITZ Adam
Cooperating Embedded Systems and Wireless Sensor Networks

MERZ Stephan, NAVET Nicolas
Modeling and Verification of Real-time Systems

PASCHOS Vangelis Th
Combinatorial Optimization and Theoretical Computer Science: Interfaces and Perspectives

WALDNER Jean-Baptiste
Nanocomputers and Swarm Intelligence

2007

BENHAMOU Frédéric, JUSSIEN Narendra, O'SULLIVAN Barry
Trends in Constraint Programming

JUSSIEN Narendra
A to Z of Sudoku

2006

BABAU Jean-Philippe *et al.*
From MDD Concepts to Experiments and Illustrations – DRES 2006

HABRIAS Henri, FRAPPIER Marc
Software Specification Methods

MURAT Cecile, PASCHOS Vangelis Th
Probabilistic Combinatorial Optimization on Graphs

PANETTO Hervé, BOUDJLIDA Nacer
Interoperability for Enterprise Software and Applications 2006 / IFAC-IFIP I-ESA'2006

2005

GÉRARD Sébastien *et al.*
Model Driven Engineering for Distributed Real Time Embedded Systems

PANETTO Hervé
Interoperability of Enterprise Software and Applications 2005

Printed and bound by CPI Group (UK) Ltd, Croydon, CR0 4YY

13/09/2026

14961016-0007